IMAGES
of America

FIREFIGHTING IN FREDERICK

IMAGES
of America

FIREFIGHTING IN FREDERICK

Clarence "Chip" Jewell
and Warren W. Jenkins

ARCADIA
PUBLISHING

ISBN 978-0-7385-1734-6

Published by Arcadia Publishing
Charleston, South Carolina

Printed in the United States of America

Library of Congress Catalog Card Number: 2004112680

For all general information contact Arcadia Publishing at:
Telephone 843-853-2070
Fax 843-853-0044
E-mail sales@arcadiapublishing.com
For customer service and orders:
Toll-Free 1-888-313-2665

Visit us on the Internet at www.arcadiapublishing.com

This book is dedicated to all firefighters from the companies in the city of Frederick who have died in the line of duty:

William Carlton   Junior Fire Company No. 2   Died December 23, 1840
(Note: First recorded Line of Duty death in the state of Maryland)
William B. Davis   Junior Fire Company No. 2   Died March 20, 1910
Walter K. Shank   Citizens Truck Company   Died November 24, 1957
Howard M. "Barney" Stroup   Citizens Truck Company   Died October 27, 1968
Gilmore "Butch" Stitely   Citizens Truck Company   Died January 10, 2001

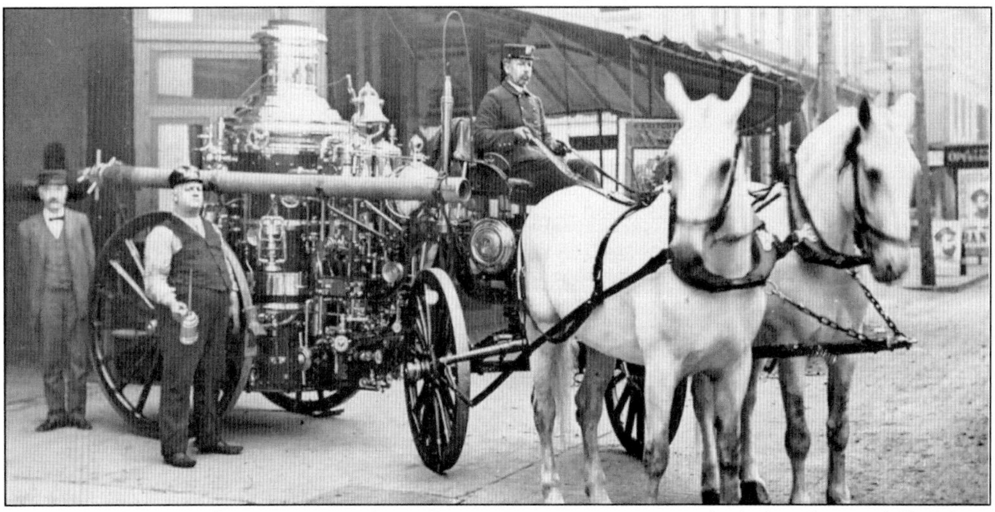

Pictured is driver William Davis on the LaFrance Junior steamer months before he was killed from the horses pulling the engine over him. The members of the Junior Fire Company were so saddened, a Christie tractor was attached to the steamer and the horses sold to Washington, D.C. Fire Department. Ironically, one of the horses was killed when it was involved in an accident in D.C.

Contents

ACKNOWLEDGMENTS

In order to make this book special, we have gathered all pictures from collections of the fire companies or individual members. We would especially like to thank John Maly, Independent Hose Company No. 1; Jack Fleischman, Independent Hose Company No. 1; Todd Johnson, Junior Fire Company No. 2; H. Richard "Rick" Hahn, Junior Fire Company No. 2; Sgt. Bill Metcalf, Junior Fire Company No. 2; Lydia Butterworth, United Steam Fire Engine Company No. 3; Paul & Rita Gordon, United Steam Fire Company No. 3; Sgt. Thomas Lawson, Citizens Truck Company; William Simmons, former owner Marken & Biefeld Printing Company; and Dawn Metcalf, Frederick County Division of Fire/Rescue Services

Warren Jenkins would like to thank Mr. and Mrs. Wayne Jenkins, David Bowen, Don French, Bob Kimball, Howard Meile, John Floyd, Kevin Ryer, Virginia McLaughlin, Mike Sanders, and the late Ed Bosanko of the Maryland State Firemen's Association.

The authors' profits from this book will be donated to the Frederick County Fire & Rescue Association to fund help the Frederick County Fire Museum and expenses for the National Fallen Firefighters Memorial Weekend.

INTRODUCTION

The following introduction is adapted from a living history presentation of the Frederick fire service that has been written and performed by the co-author of this book, Clarence "Chip" Jewell.

Hear ye, Hear ye!! All able bodied men of Fredericktowne are hereby summoned to organize a fire company to protect our city from fire and conflagration.

The year is 1818, and the citizens of Frederick have gathered around the courthouse square to discuss the fire protection, or shall we say, the lack thereof, that existed in our city. There were fire companies indeed—the Union Fire Company, the Friendship Fire Company, the Sun Fire Company, and the Frederick Hose Company—but their abilities and capabilities were somewhat in question.

The fire service in Frederick can actually be traced back to 1760, when the town fathers petitioned the Maryland Assembly for permission to conduct a lottery to purchase a fire engine, which they did in 1764. This engine, known as "Grandfather," was stationed with the Frederick Hose Company, predecessor to the Independent Hose Company No. 1. Of course, the Independents are the oldest continually operating fire company in the state of Maryland, having been reorganized from the Frederick Hose Company in 1818. In 1817, the Town of Frederick passed an ordinance requiring "all white males, except men of the Gospel, to respond with bucket in hand" or be fined the princely sum of 12.5¢.

In 1838, a disastrous fire occurred on South Street, destroying many structures. The antiquated equipment of the day led a group of young men to gather at the drug store on the square corner to begin discussion of a new fire company. These men gathered at the Dill House to organize the Young Men's Fire Company. When the company took delivery of the John Rodgers Junior engine, the company ultimately organized as the Junior Fire Company.

Tragedy struck the Junior Fire Company, and the entire state of Maryland, in 1840 when William Carlton, a founding member of the Juniors and city solicitor, collapsed in the street while pulling the Junior pumper through the streets of Frederick in response to a fire near the Court House Square. Thus, he became the first line of duty death recorded in the state of Maryland.

In 1837, the Washington Hose Company was organized in the south end of Frederick in the area of West All Saints Street, but it faltered and was disbanded.

Many former members of the Washington Hose Company gathered to reorganize, at first calling themselves the Mechanicks Hose Company but later changing the name to the United Hose Company. The city provided $430 towards the purchase of land to erect a firehouse. This location was in a swampy area adjoining Carroll Creek. Thus, the fire company members were dubbed "Swampers" and the firehouse called the "Swamp Hall." To this day, you can pass the firehouse and see a man called "Johnny Swamper" atop the cupola.

In 1858, Capt. John Sinn of the United Fire Company suggested the Uniteds form a home guard militia. The Uniteds formed the United Guard, followed by the Independent Riflemen and the Junior Defenders.

In 1857, a fire was smoldering in our nation's heartland and required the assistance of the Frederick volunteers. Dred Scott, a former Missouri slave who had lived 11 years as a free man in the state of Illinois and the Wisconsin Territory, returned to his home state of Missouri only to once again be declared a slave. He appealed the decision through the courts, all the way to the Supreme Court of the United States, where a Frederick lawyer, the brother-in-law of Francis Scott Key, Chief Justice Roger Brooke Taney, wrote what was to become known as the Dred Scott Decision.

On October 17, 1859, at 10:00 a.m., the alarm bell rang atop the United Fire Hall, followed shortly by the bells of the Independents and Juniors, summoning the volunteers to their stations, not to fight a fire but to help quell an insurrection. The volunteers rushed to their stations and marched to the Baltimore and Ohio Railroad Depot, where they boarded a special train. They were then transported to Harpers Ferry and federalized to help the troops.

The Civil War broke out and directly involved the fire companies and their militias. Units were assigned to guard railroad bridges in Frederick and at Harpers Ferry. Fire houses were used as arsenals. The Junior Fire Hall was also used as a part of General Hospital #6 and received wounded from the Battle of Antietam. It should be noted that the Junior Fire Hall was also the headquarters of the local Sons of Temperance. It is unlikely that any fire company could accommodate that organization today.

The 1870s saw the era of steam in the Frederick fire service. The first steam fire engine was purchased by the Junior Fire Company No. 2 in 1876. This engine was a Silsby steam engine and served until 1908. The Juniors also purchased the first ladder wagon, a unit built by the Charles T. Holloway Company of Baltimore.

The Uniteds soon followed by purchasing a Clapp & Jones engine that became one of the most famous steam pumpers in the country. The engine arrived under the cover of darkness at the B&O Railroad yard off Carroll Street. Late at night, members removed the cover to unload the engine; however, instead of a shiny new fire engine, they found a large, black metal monstrosity. They were so disappointed and upset that they summoned Mr. Clapp to Frederick. Upon his arrival, he directed the volunteers of the Uniteds to pour boiling water over the engine to remove the protective covering that had been placed over it for shipment. The unit was unloaded and the next day paraded throughout the streets of Frederick.

Mr. Clapp was so confident in his engine that he hand wrote on the contract that it would throw two streams of water over the town clock, which it did. This engine was nicknamed the "Lily of the Swamp" and later donated to the Smithsonian Institute, where it was on display for many years.

Not to be outdone, in 1880, the Independent Hose Company purchased a LaFrance steam pumper called "Romeo." It, however, met an untimely demise. While responding to a fire, Romeo crashed through a bridge over Carroll Creek and was completely destroyed.

Undaunted, the Independents contacted the LaFrance Engine Company and had a new engine built to replace Romeo. Named "Juliet," this unit was a fine piece of equipment, pumping 2,800 feet of hose at the Frederick Fairgrounds to fight a barn fire. In 1884, it was transported by railcar 17 miles in 18 minutes to Emmitsburg where it pumped several days in sub-freezing temperatures to fight a fire at St. Joseph's College to save what we know today as the National Fire Academy.

The year 1893 saw much to celebrate in Frederick. The Independent Hose Company purchased an ornate hose wagon from the Charles T. Holloway Company of Baltimore. That year also saw the organization of the Maryland State Firemen's Association. Mr. J. Roger McSherry of the Independent Hose Company was elected the first president, with Mr. Charles T. Holloway of the Baltimore Volunteer Firemen's Veterans serving as the vice president.

One

INDEPENDENT HOSE COMPANY NO. 1

Oldest Fire Company in the State of Maryland

The Independent Hose Company No. 1 has been in continuous operation since 1818, making it the oldest fire company in the state of Maryland and one of the oldest in the country.

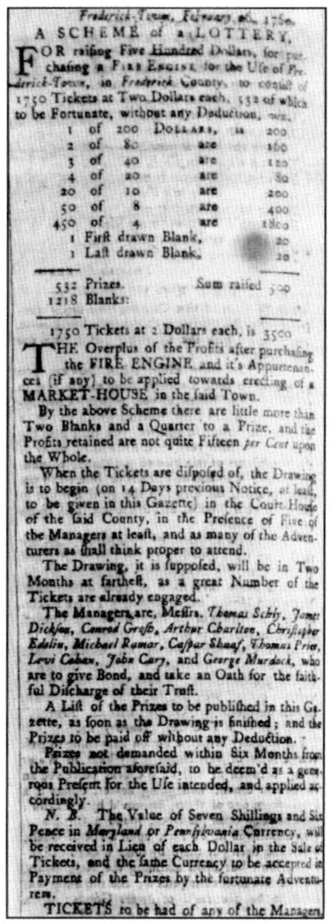

Frederick-Town, February [illegible]

A SCHEME of a LOTTERY,

FOR raising Five Hundred Dollars, for purchasing a FIRE ENGINE for the Use of *Frederick-Town*, in *Frederick* County, to consist of 1750 Tickets at Two Dollars each, 532 of which to be Fortunate, without any Deduction, viz.

1	of	200	DOLLARS, is	200
2	of	80	are	160
3	of	40	are	120
4	of	20	are	80
20	of	10	are	200
50	of	8	are	400
450	of	4	are	1800
1	First drawn Blank,			20
1	Last drawn Blank,			20
532	Prizes.		Sum raised	500
1218	Blanks:			
1750	Tickets at 2 Dollars each, is			3500

THE Overplus of the Profits after purchasing the FIRE ENGINE and it's Appurtenances (if any) to be applied towards erecting of a MARKET-HOUSE in the said Town.

By the above Scheme there are little more than Two Blanks and a Quarter to a Prize, and the Profits retained are not quite Fifteen *per Cent* upon the Whole.

When the Tickets are disposed of, the Drawing is to begin (on 14 Days previous Notice, at least, to be given in this Gazette) in the Court-House of the said County, in the Presence of Five of the Managers at least, and as many of the Adventurers as shall think proper to attend.

The Drawing, it is supposed, will be in Two Months at farthest, as a great Number of the Tickets are already engaged.

The Managers are, Messrs. *Thomas Schly*, *James Dickson*, *Conrad Grosh*, *Arthur Charlton*, *Christopher Edelin*, *Michael Ramar*, *Caspar Shaaf*, *Thomas Price*, *Levi Cohen*, *John Cary*, and *George Murdock*, who are to give Bond, and take an Oath for the faithful Discharge of their Trust.

A List of the Prizes to be published in this Gazette, as soon as the Drawing is finished; and the Prizes to be paid off without any Deduction.

Prizes not demanded within Six Months from the Publication aforesaid, to be deem'd as a generous Present for the Use intended, and applied accordingly.

N. B. The Value of Seven Shillings and Six Pence in *Maryland* or *Pennsylvania* Currency, will be received in Lieu of each Dollar in the Sale of Tickets, and the same Currency to be accepted in Payment of the Prizes by the fortunate Adventurers.

TICKETS to be had of any of the Managers.

In 1764, the town fathers of Frederick petitioned the Maryland Assembly for permission to conduct a lottery, part of the proceeds to purchase Frederick's first fire engine, known as "Grandfather." Grandfather was assigned to the Frederick Hose Company, predecessor to the Independent Hose Company No. 1, that same year. (Courtesy Paul and Rita Gordon.)

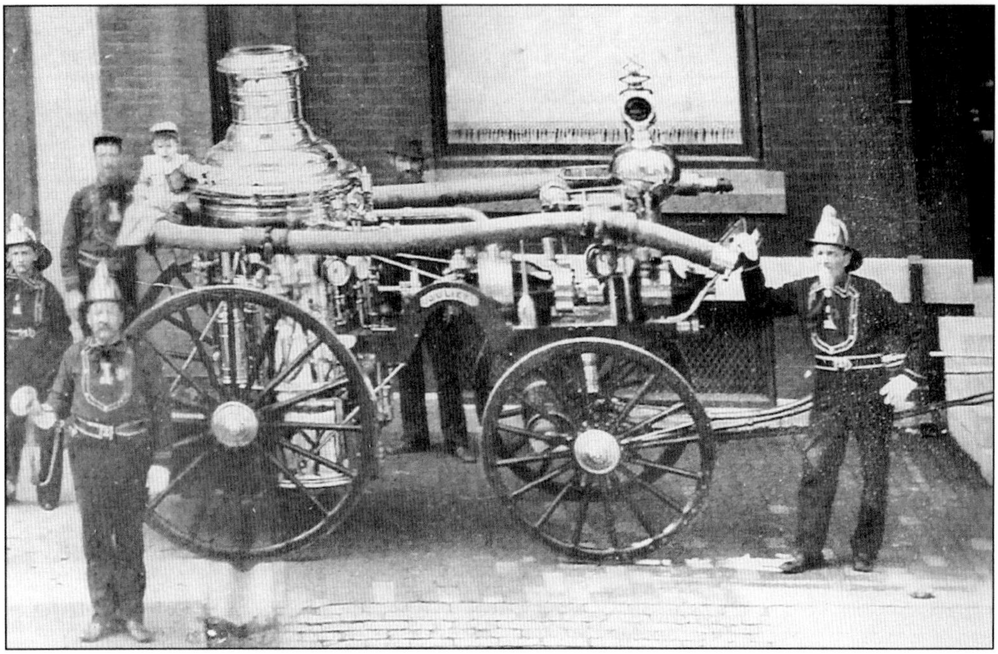

The first steam pumper purchased by the Independents was a LaFrance pumper named "Romeo." Romeo was destroyed in 1881 responding to an alarm of fire when it fell through a bridge over Carroll Creek. Shown is "Juliet," the duplicate pumper purchased to replace the doomed engine. Juliet is credited with saving the buildings of St. Joseph College in Emmitsburg in 1885. Today, the campus is home of the National Fire Academy. (Courtesy IHC [Independent Hose Company No. 1].)

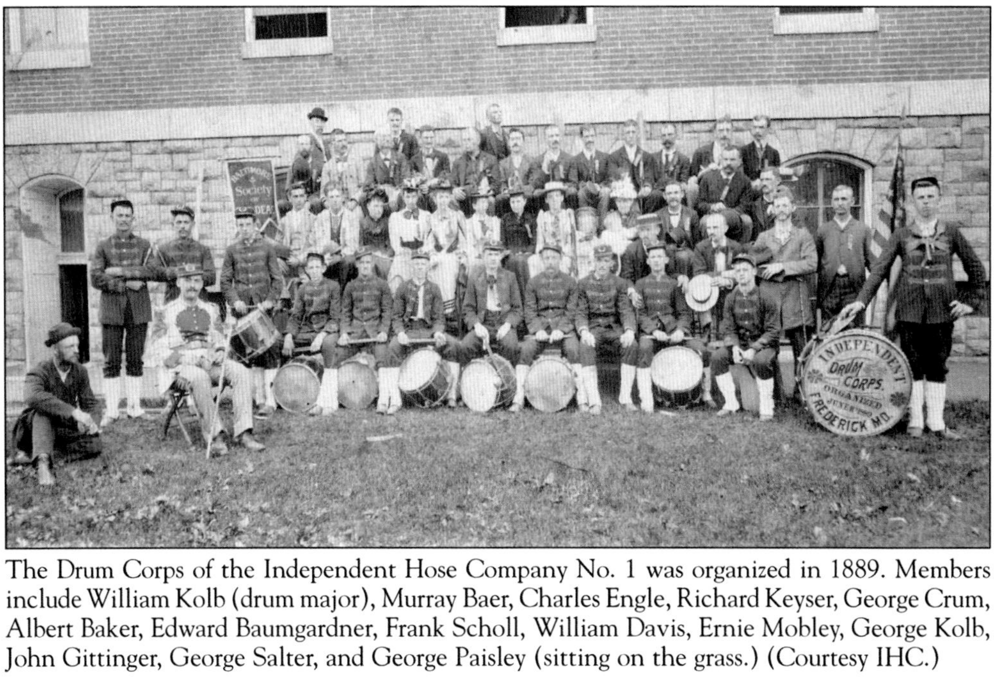

The Drum Corps of the Independent Hose Company No. 1 was organized in 1889. Members include William Kolb (drum major), Murray Baer, Charles Engle, Richard Keyser, George Crum, Albert Baker, Edward Baumgardner, Frank Scholl, William Davis, Ernie Mobley, George Kolb, John Gittinger, George Salter, and George Paisley (sitting on the grass.) (Courtesy IHC.)

This photo shows some early uniforms of the Independents. Note the smokers with cigars and pipe. (Courtesy IHC.)

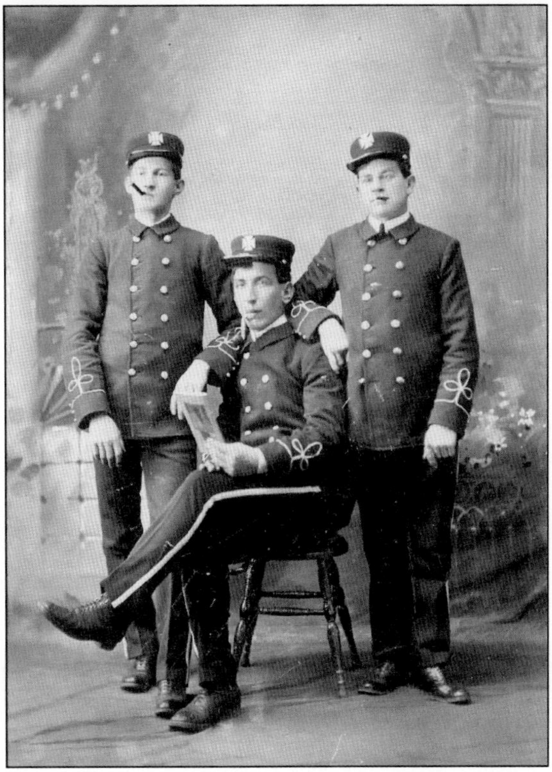

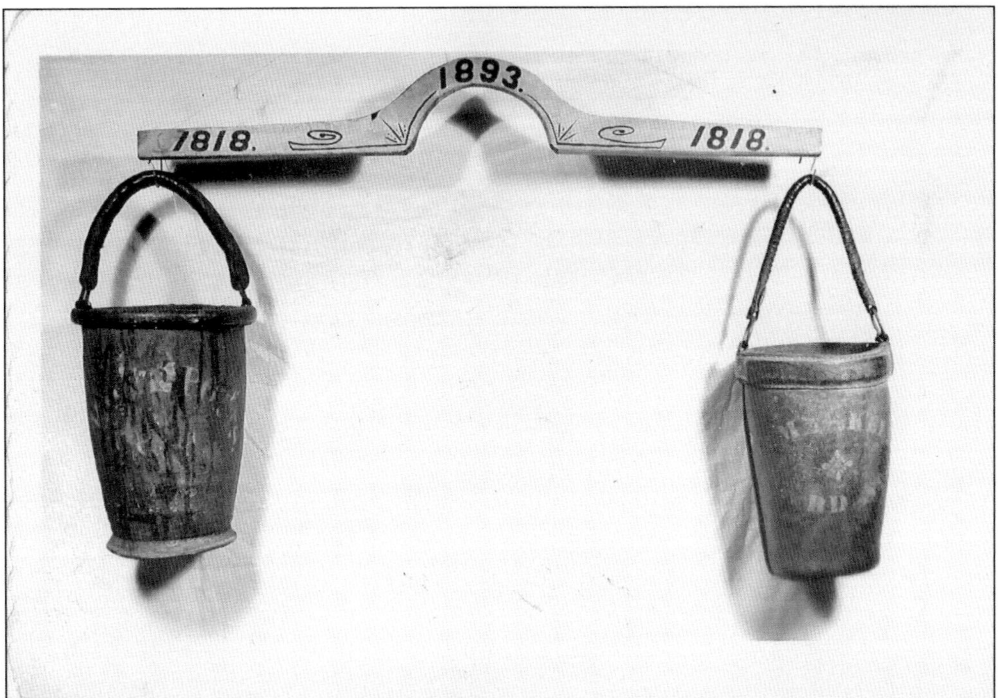

Leather buckets were used to supply early pumping engines. A firefighter could carry two buckets with the carrier shown in the picture. (Courtesy IHC.)

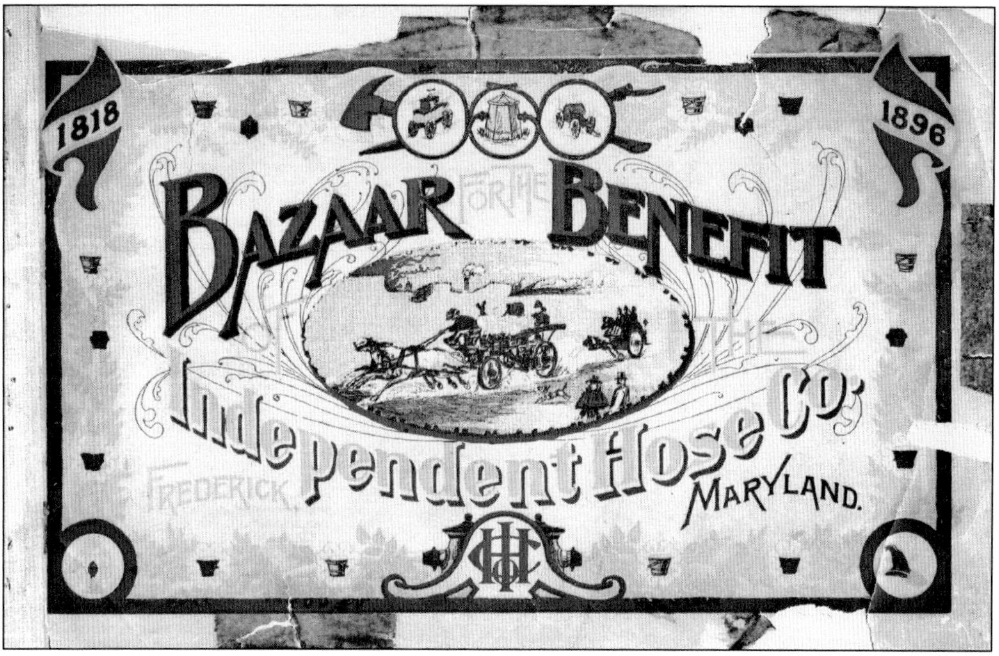

A benefit bazaar was held by the Independent Hose Company No 1 in 1896. Many of the bazaars were held at the City Opera House on North Market Street. (Courtesy IHC.)

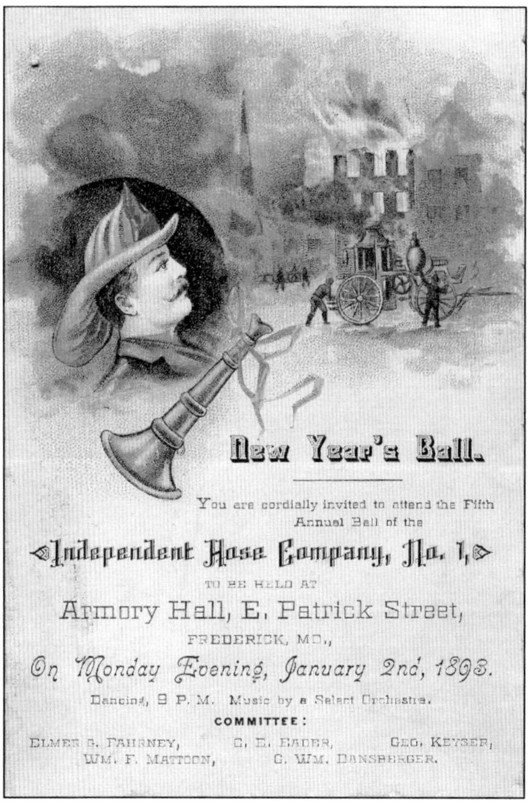

This is a dance card from the New Year's Ball in 1898, which was held at the Armory Hall to benefit the Independents. (Courtesy IHC.)

This dress uniform was worn for the first parade of the drum corps. The uniform included a pasteboard green beaver hat containing a white "1" rather than the helmet shown. (Courtesy IHC.)

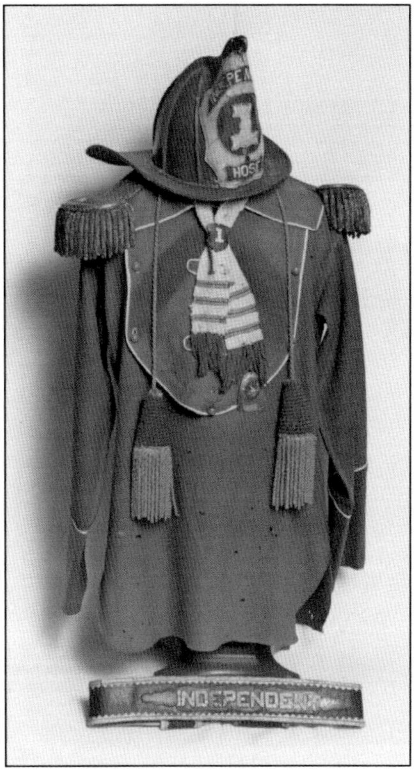

Independent Hose Company No. 1 also housed the Odd Fellow's Hall. This 1868 picture may have been taken to celebrate the 50th anniversary of the fire company. (Courtesy IHC.)

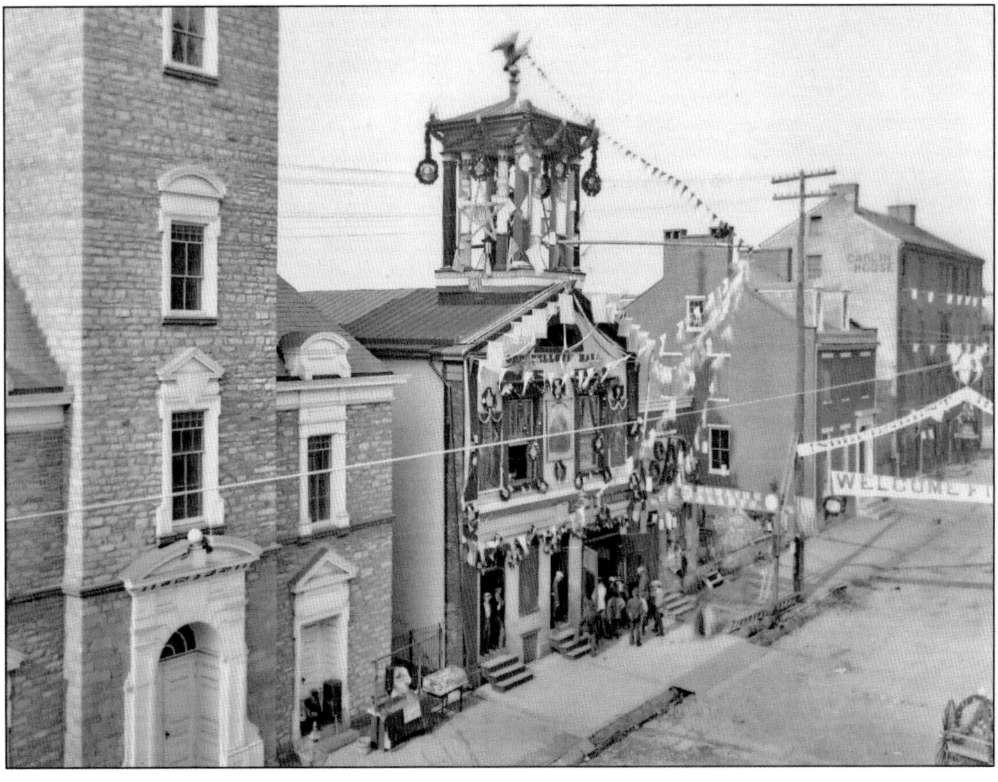

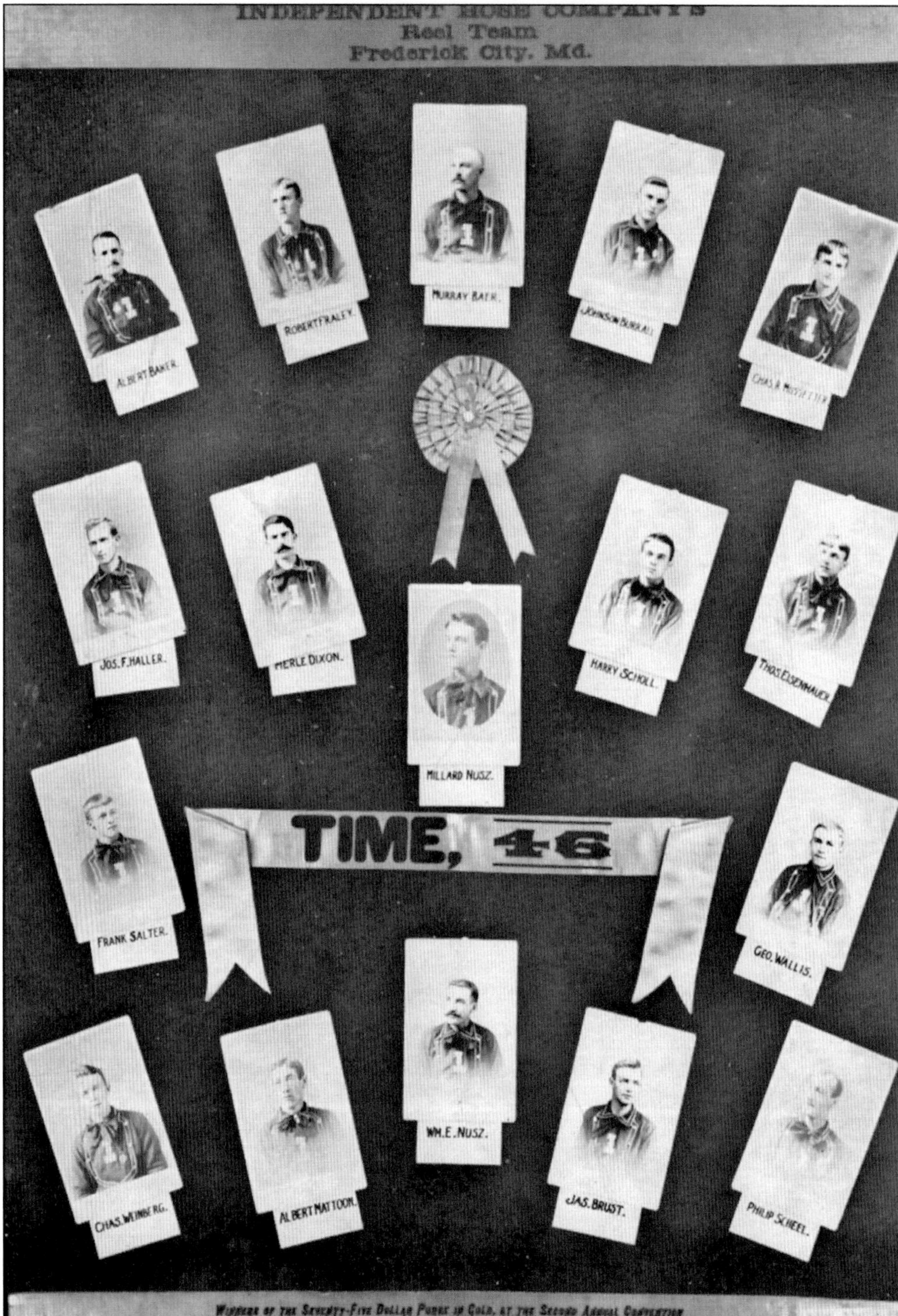

The Independents were the winning hose reel team at a competition in Hagerstown, Maryland, part of the second Maryland State Firemen's Association Convention, in 1894. (Courtesy IHC.)

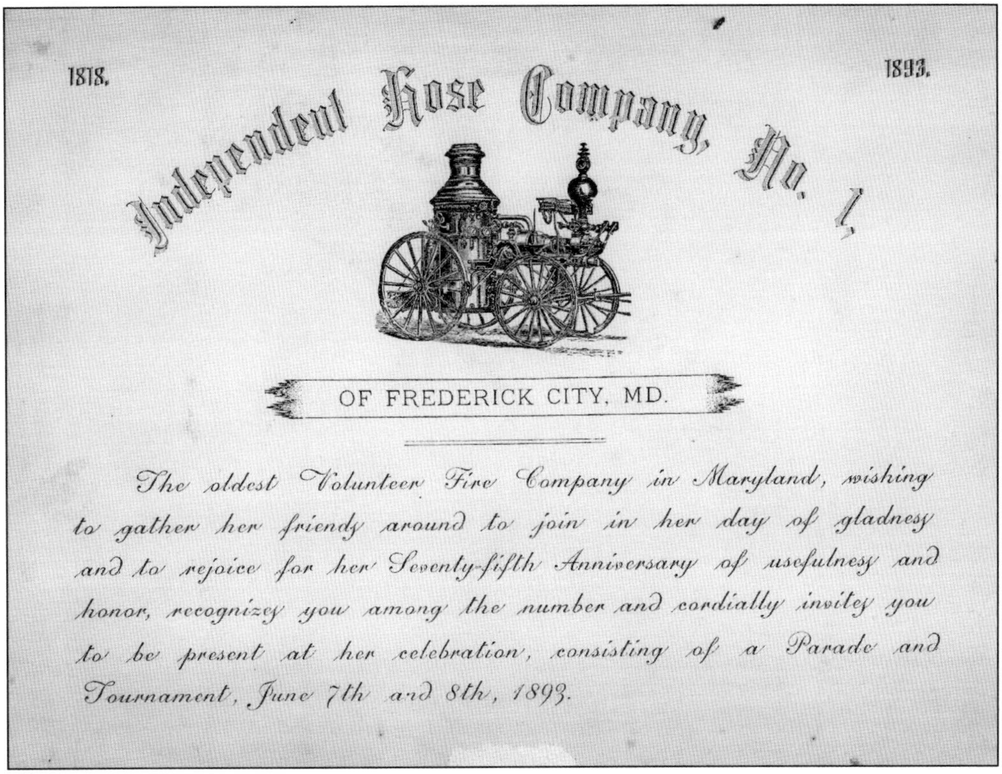

1818. 1893.

Independent Hose Company, No. 1

OF FREDERICK CITY, MD.

The oldest Volunteer Fire Company in Maryland, wishing to gather her friends around to join in her day of gladness and to rejoice for her Seventy-fifth Anniversary of usefulness and honor, recognizes you among the number and cordially invites you to be present at her celebration, consisting of a Parade and Tournament, June 7th and 8th, 1893.

Shown here is an invitation to attend the 75th Anniversary Parade and Tournament of the Independent Hose Company No. 1. (Courtesy IHC.)

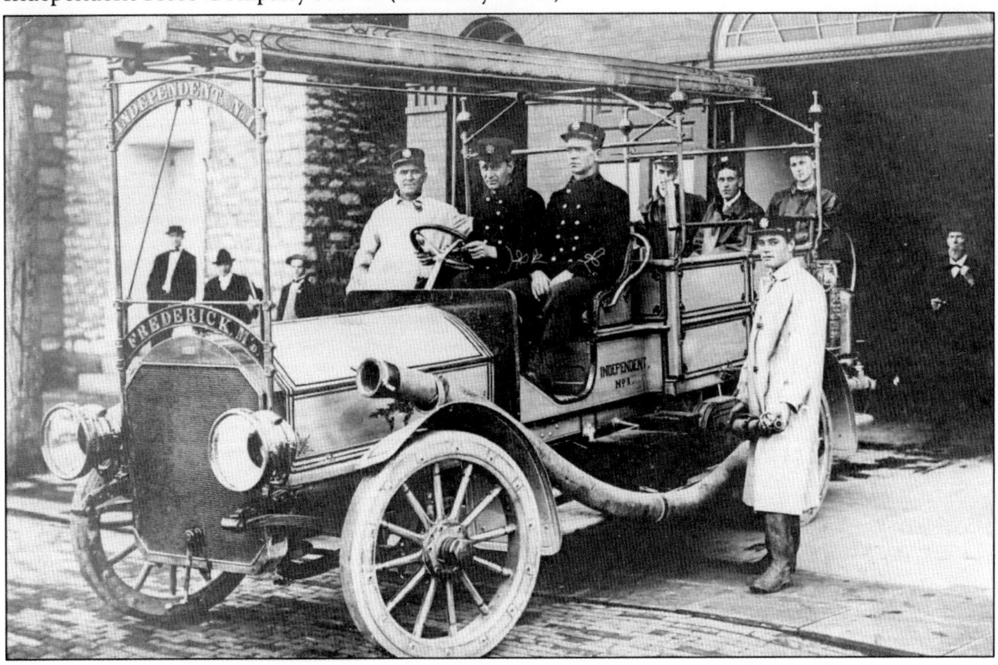

The Independent Hose Company No. 1 purchased the first automobile-type fire apparatus in Maryland. Members of the company proudly display the 1908 Howe pumper. (Courtesy IHC.)

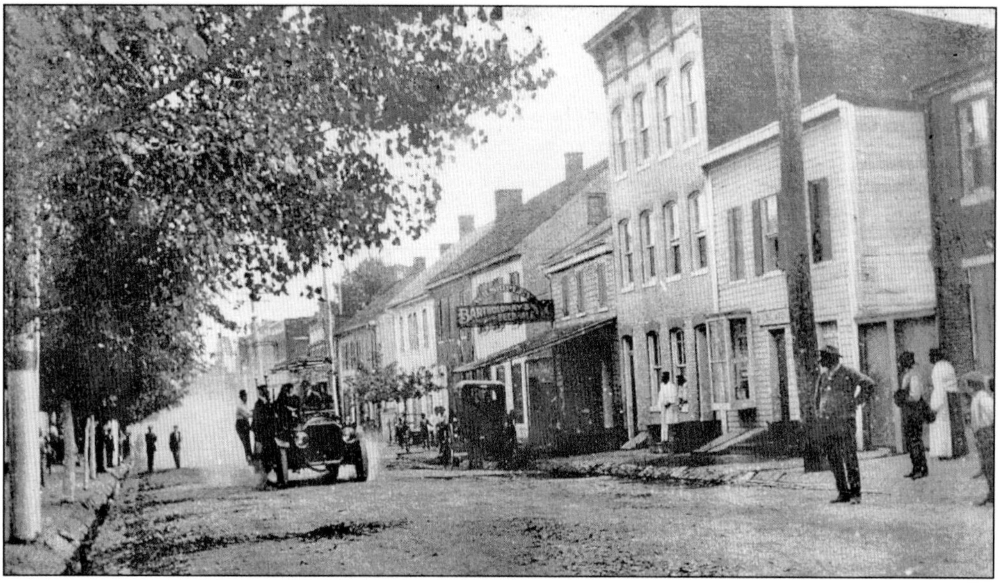

Neighbors run to see the "new fangled automobile fire engine" race to an alarm on North Market Street in Frederick. (Courtesy Paul and Rita Gordon.)

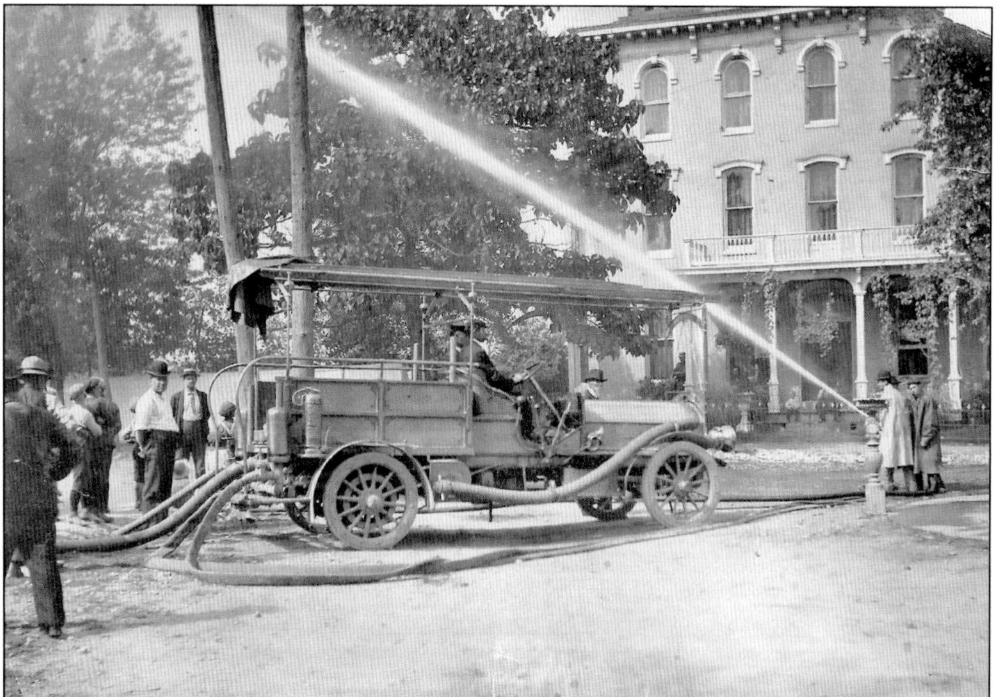

The 1908 Howe pumps several lines at a demonstration at the town fountain on North Market Street. (Courtesy IHC.)

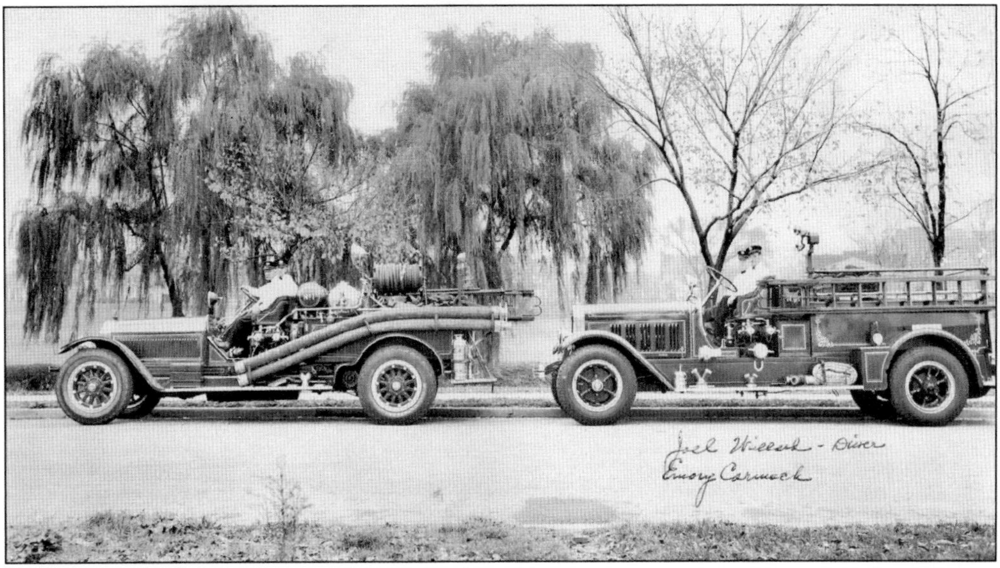

Posed next to Baker Park in Frederick are the 1921 American LaFrance on the left and the 1931 American LaFrance, manned by Joel Willard and Emory Carmack, on the right. (Courtesy IHC.)

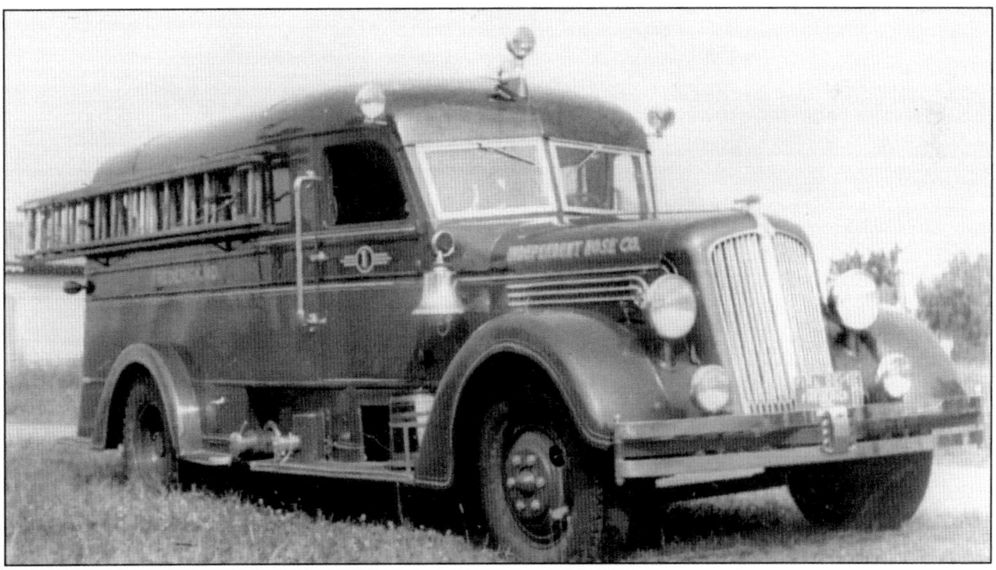

This 1937 Sedan Cab Seagrave engine was painted a dark green, the company color. This unit remained in service long after the 1954 retirement, when it was sold to the new Carroll Manor Volunteer Fire Company, which members of the Independents helped to organize. (Courtesy IHC.)

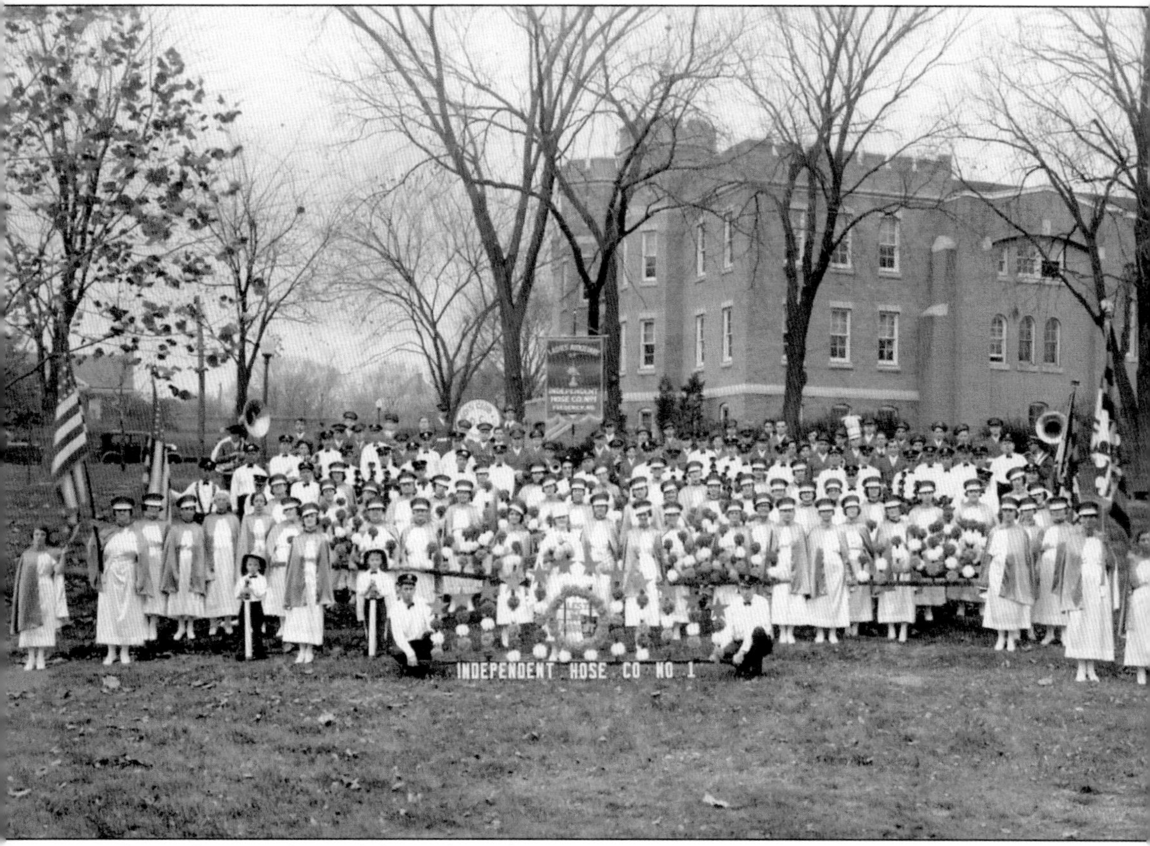

The Ladies Auxiliary of the Independent Hose Company No. 1 are shown in uniform with a local band in front of the armory on North Bentz Street. The two small children are twins Charles "Buck" Wisner and John E. Wisner. Both men went on to become active members of the Independent Hose Company No. 1 and were awarded life memberships for their service to the fire company. (Courtesy IHC.)

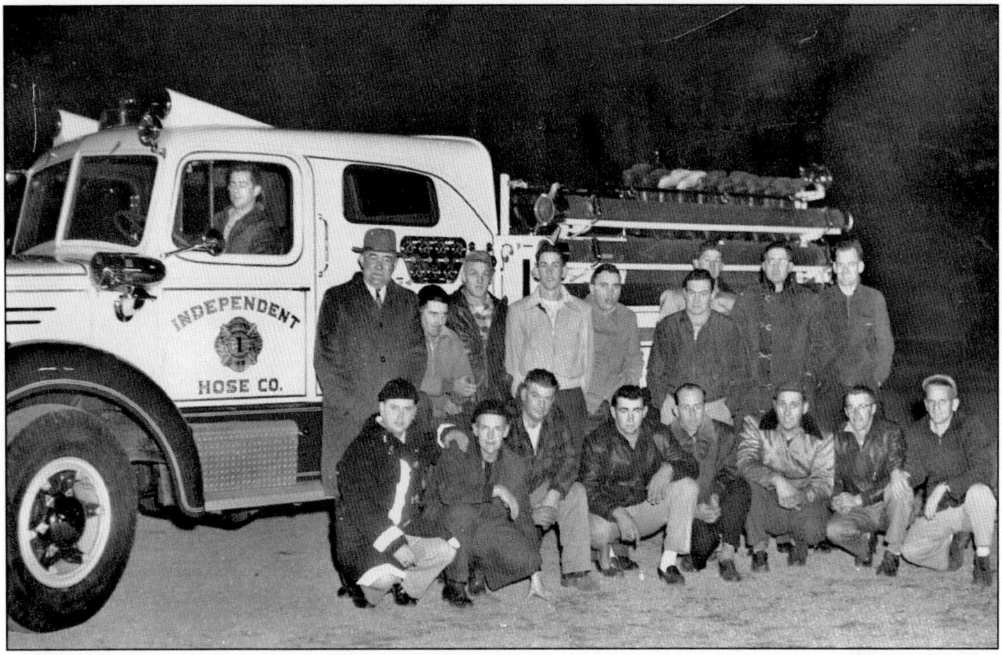

Members of the newly organized Carroll Manor Fire Company gather next to the new Mack pumper of the Independents during a fire training class. Driving for the Independents was Frank Abrecht. (Courtesy Carroll Manor Volunteer Fire Company.)

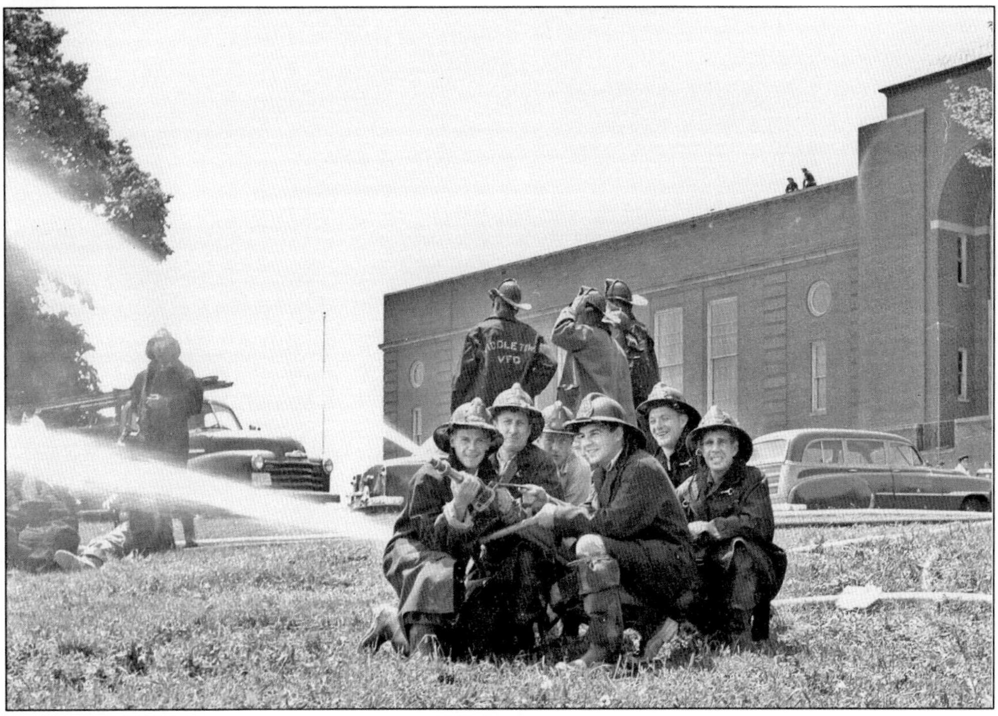

Members of the Independents practice operating a nozzle at a drill at Frederick High School during a training class conducted by the University of Maryland Fire Service Extension. (Courtesy IHC.)

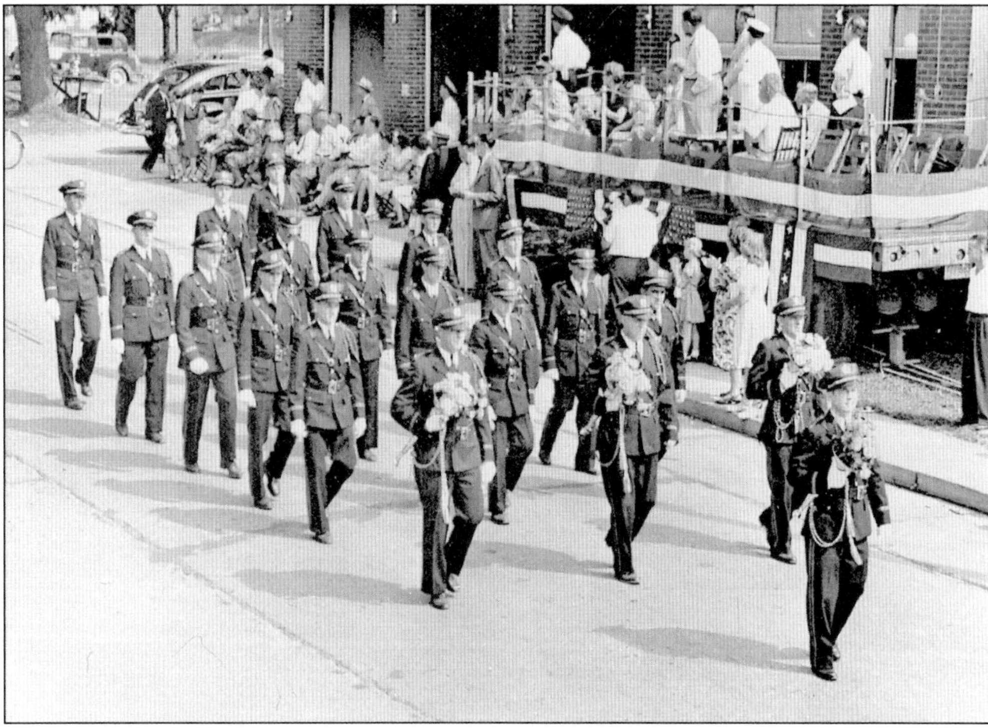

Members of the Uniform Rank of the Independent Hose Company No. 1 march in a local parade in the late 1940s. The Uniform Rank was a separate organization within the fire company to promote funds for fellowship and parade attendance. (Courtesy IHC.)

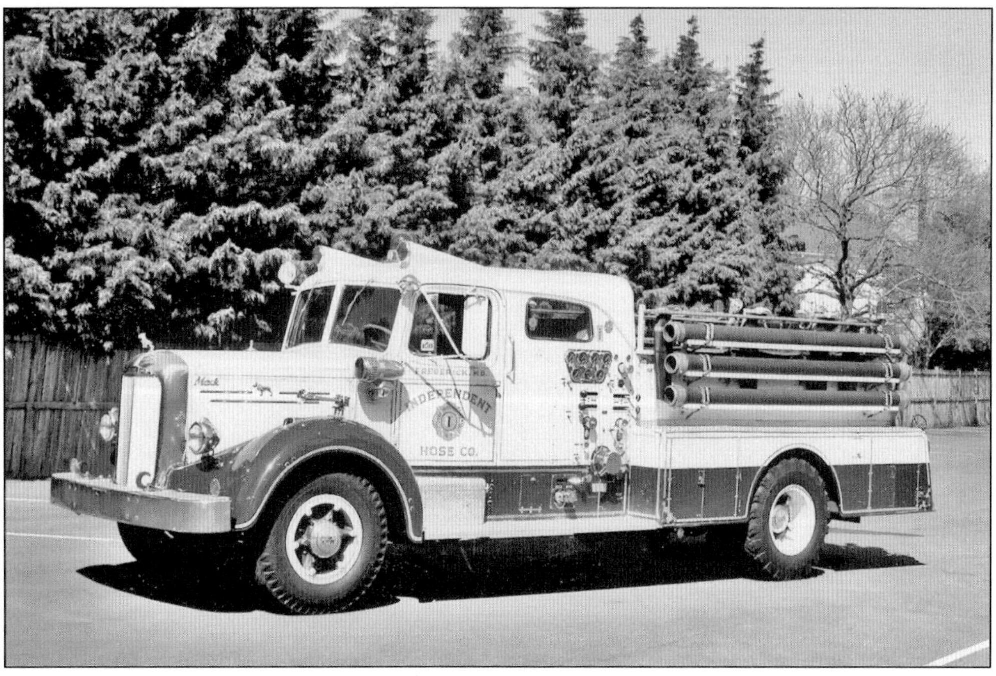

The 1953 Mack "city engine" was purchased by the City of Frederick to respond to city calls. Dressed in green and white, it was the pride of the Independents. (Courtesy Warren Jenkins.)

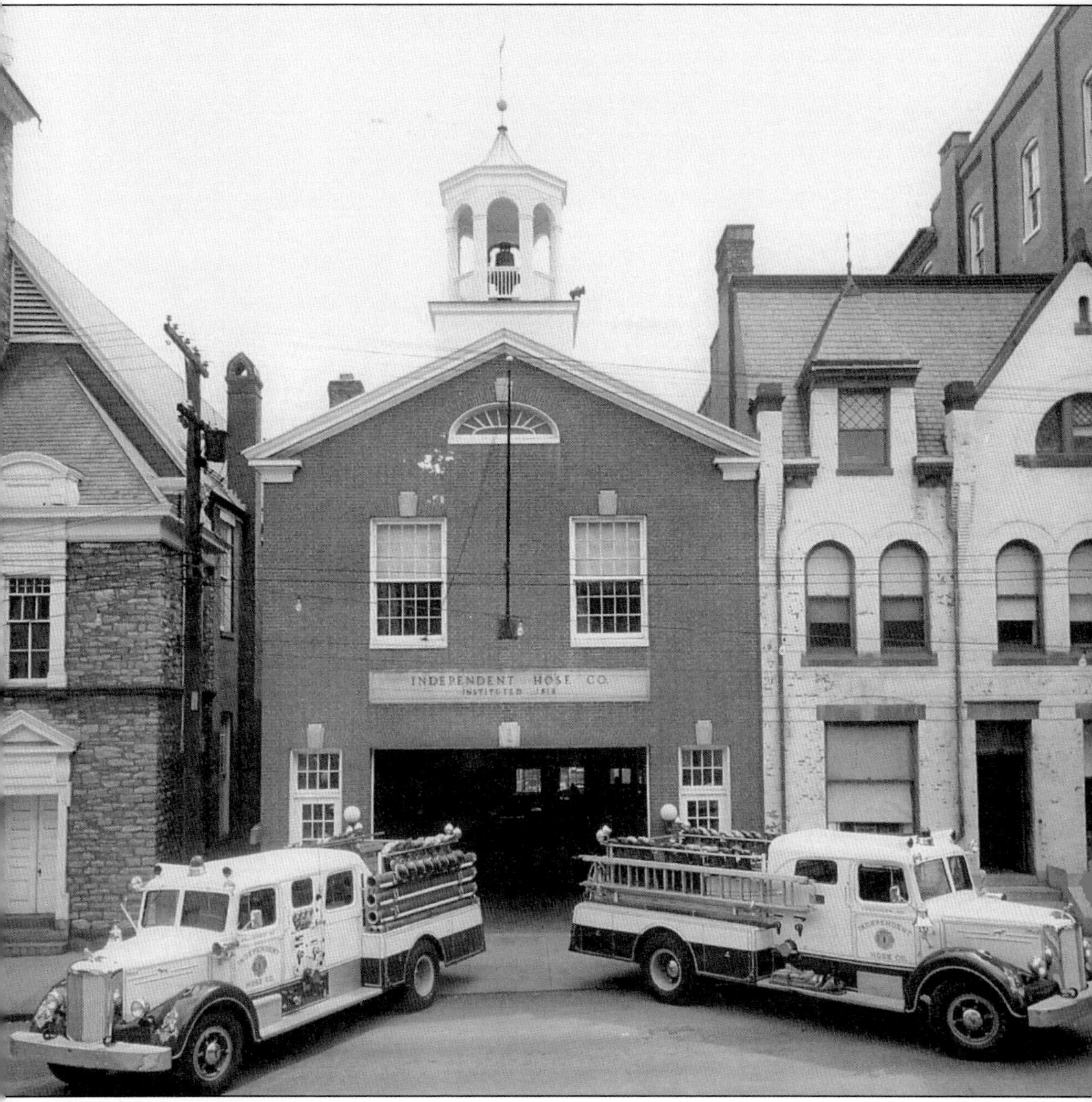

The 1953 and 1954 Mack pumpers are proudly displayed in front of the quarters on West Church Street in downtown Frederick. (Courtesy IHC.)

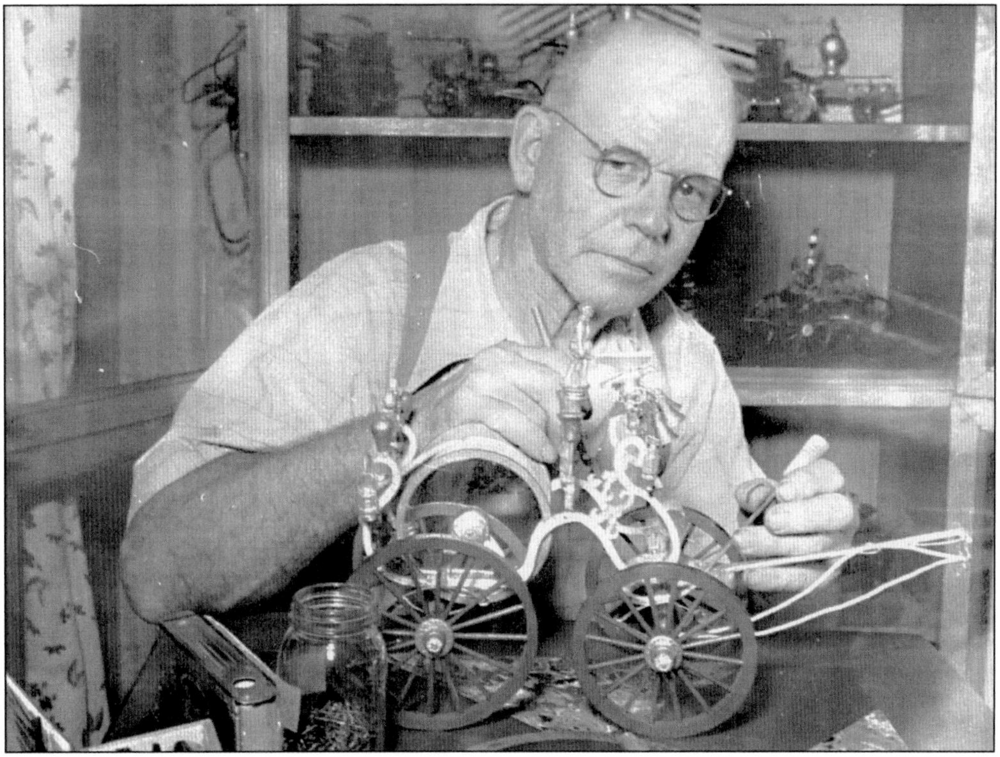

Mr. Walter Danner, member for over 35 years, handcrafted many wooden models of fire equipment of the Frederick area. (Courtesy IHC.)

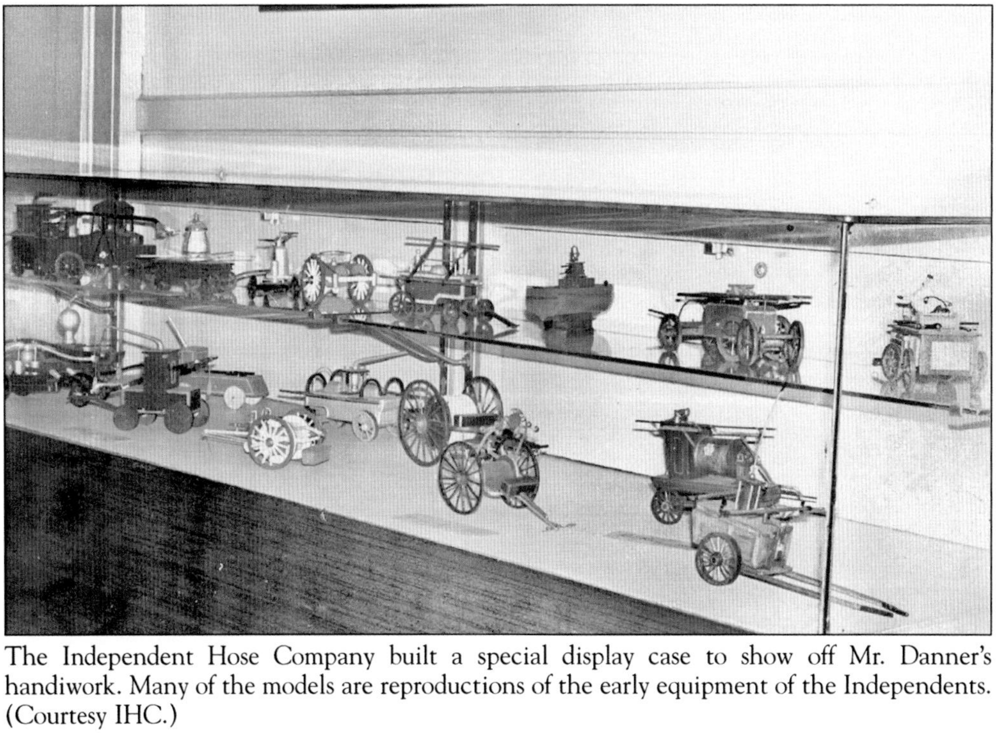

The Independent Hose Company built a special display case to show off Mr. Danner's handiwork. Many of the models are reproductions of the early equipment of the Independents. (Courtesy IHC.)

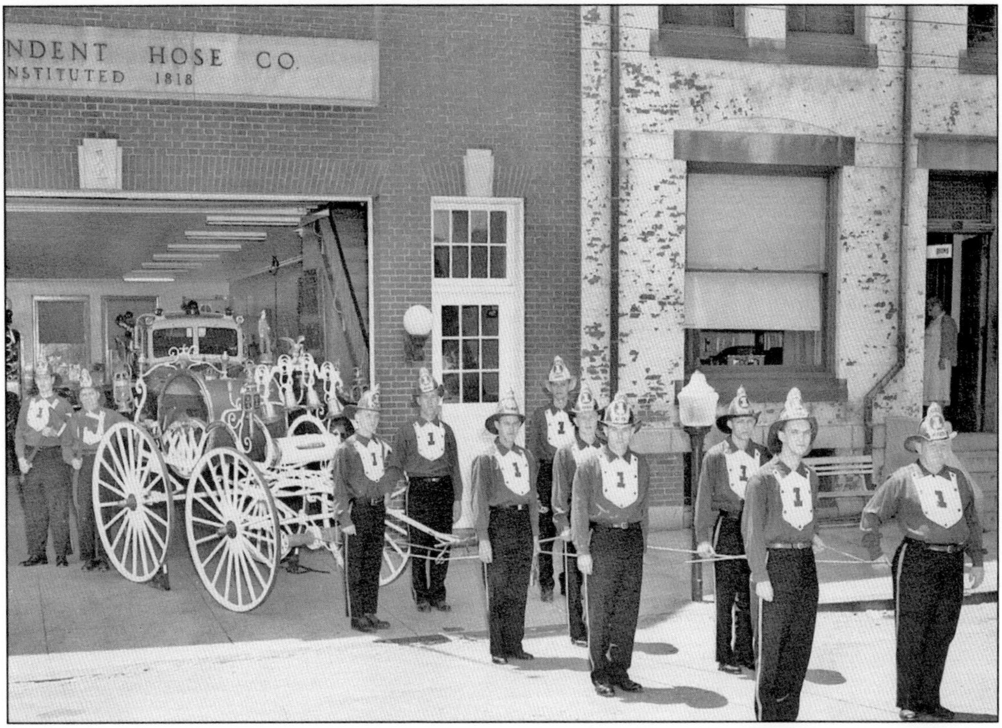

"Juliet," the 1893 Holloway Hose Reel, is in full ornamentation at the Church Street location. Members include Seymore Carbaugh, Richard Morgan, James "Dink" Rinehart, Lewis Leatherman, Paul Eigenbrode, Jack Fleischman, Bill Shaff, Charles Handley, Jerry Quill, James Lacy, and Thomas Partis. (Courtesy IHC.)

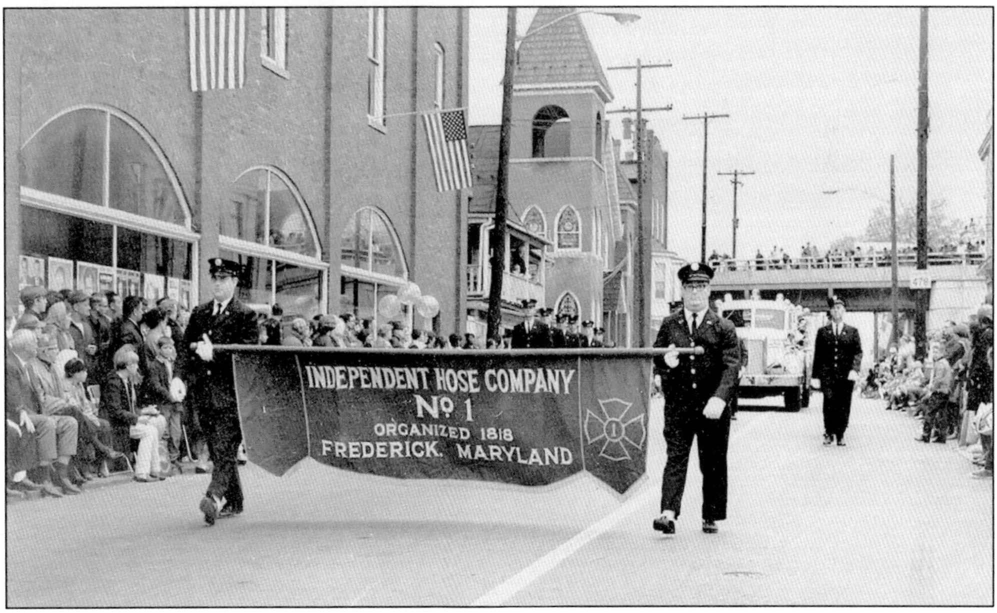

The marching unit and equipment of the Independents participate in the annual Veteran's Day Parade in Brunswick, Maryland, in the mid-1950s. Shown from left to right are Randy Baker, Jack Fleischman, Gus Draper, and Charles Handley.

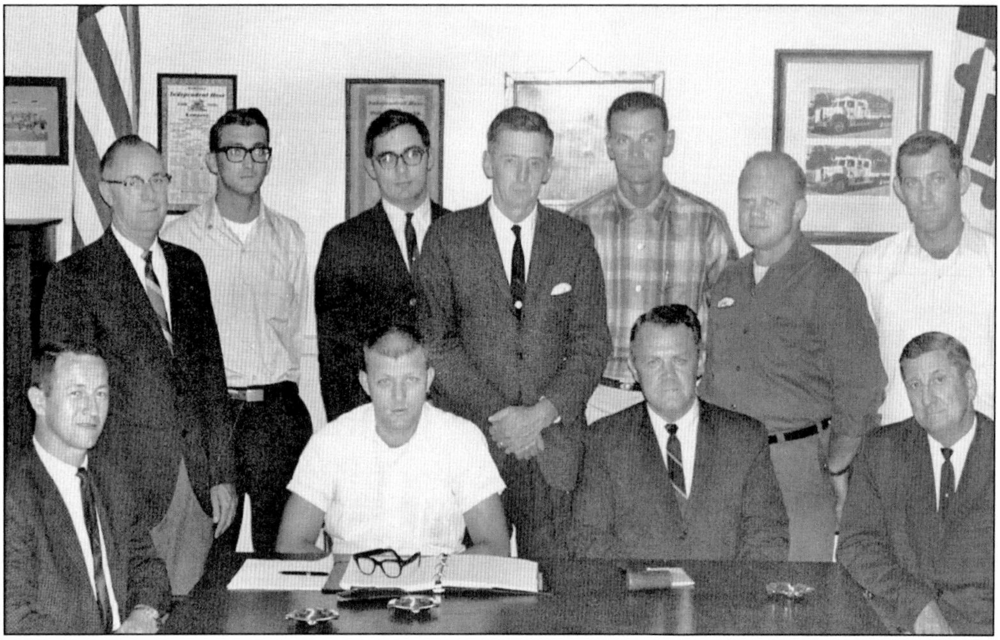

The 1969 administrative officers of the Independent Hose Company No. 1 are, from left to right, (seated) Thomas Angleberger, Dink Rinehart, Willard Horine, and John Kennedy; (standing) Melvin Schwearing, Hiley Orndorff, Jim Devilbiss, Lester Baker, Jack Fleischman, Chester Stevens, and Bob Rinehart. (Courtesy IHC.)

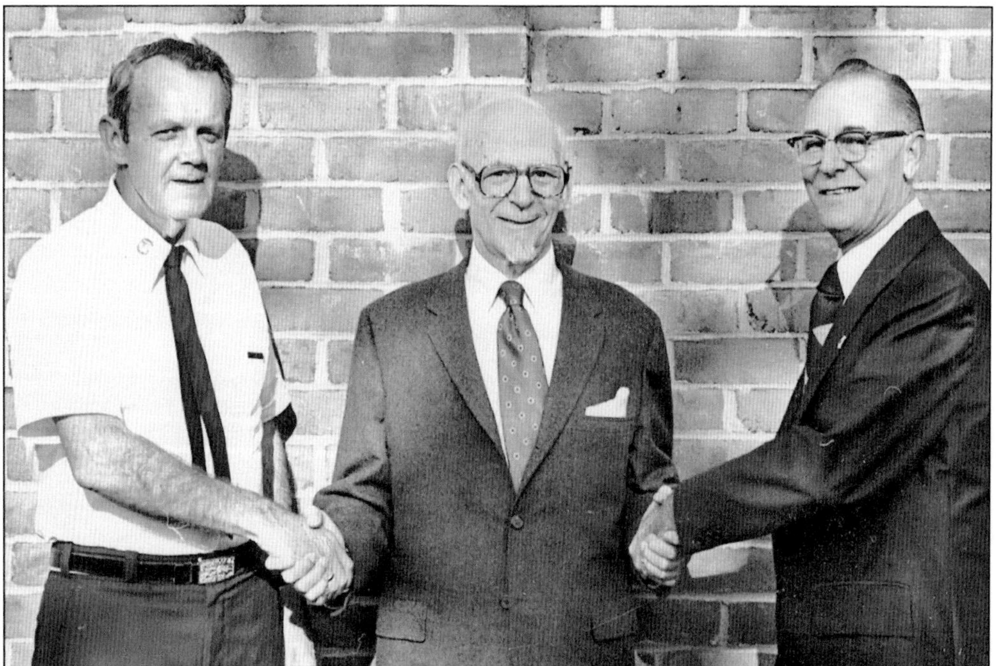

Three presidents of the Independent Hose Company, from left to right, are Willard Horine, 1969–1977; Benjamin B. Rosenstock, 1939–1969; and Melvin Schwearing, 1978. (Courtesy IHC.)

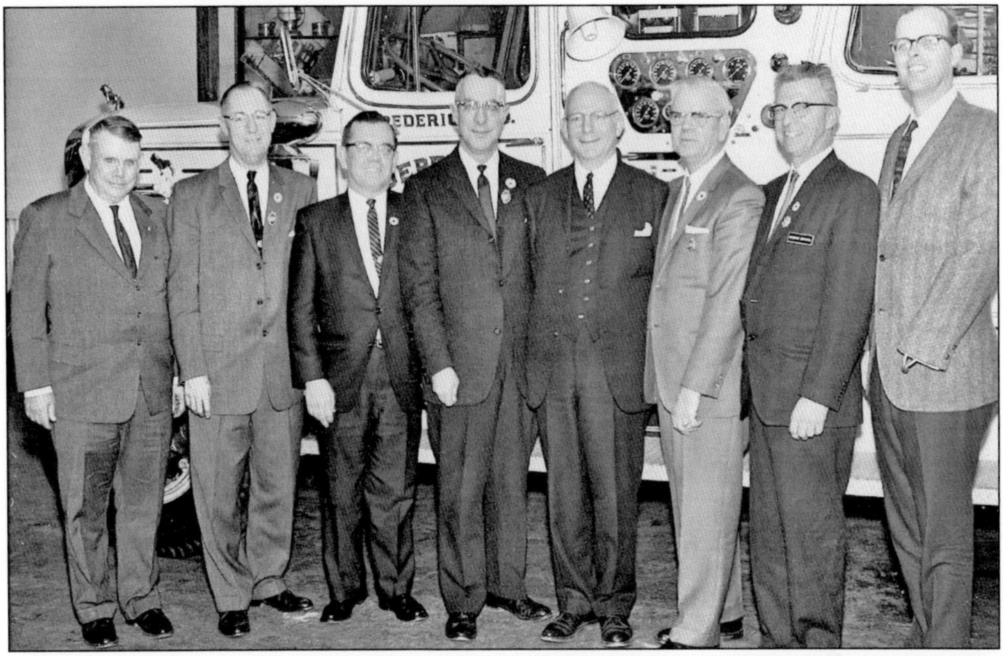

Members of the executive committee of the Maryland State Firemen's Association met in Frederick during the term of Pres. William Moore, from 1967 to 1968. Moore was a Frederick native and member of the United Steam Fire Engine Company No. 3 and Citizens Truck Company. (Courtesy IHC.)

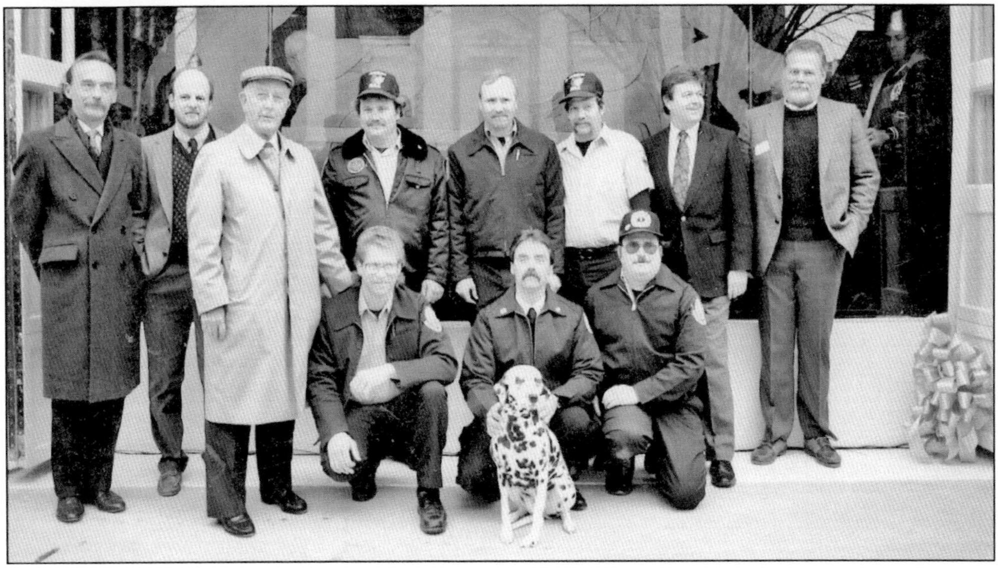

The former Independent fire house in downtown Frederick was converted to an office center, named the Firehouse Financial Center. A large glass etching was erected across the former engine bay depicting firefighters responding to a call. Members of the Independent Hose Company who were used as models for the etching attended the dedication. From left to right are (standing) two unidentified building representatives, Gov. William Donald Shaffer, Chuck Handley, Dewey Foreman, John Grice, Mayor Ron Young, and an unidentified building representative; (kneeling) Bob Page, Bill Horine, and Bob Stevens. (Courtesy IHC.)

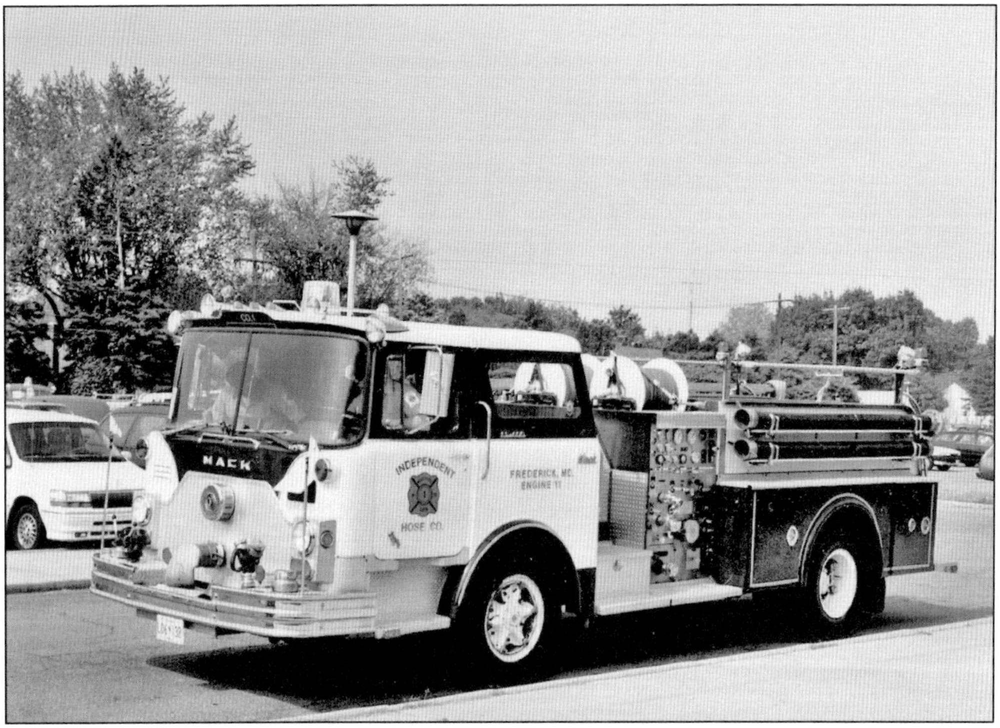

Former Engine 11 is a 1973 CF Mack restored and owned by Independent member Ed Crutchley. (Courtesy Warren Jenkins.)

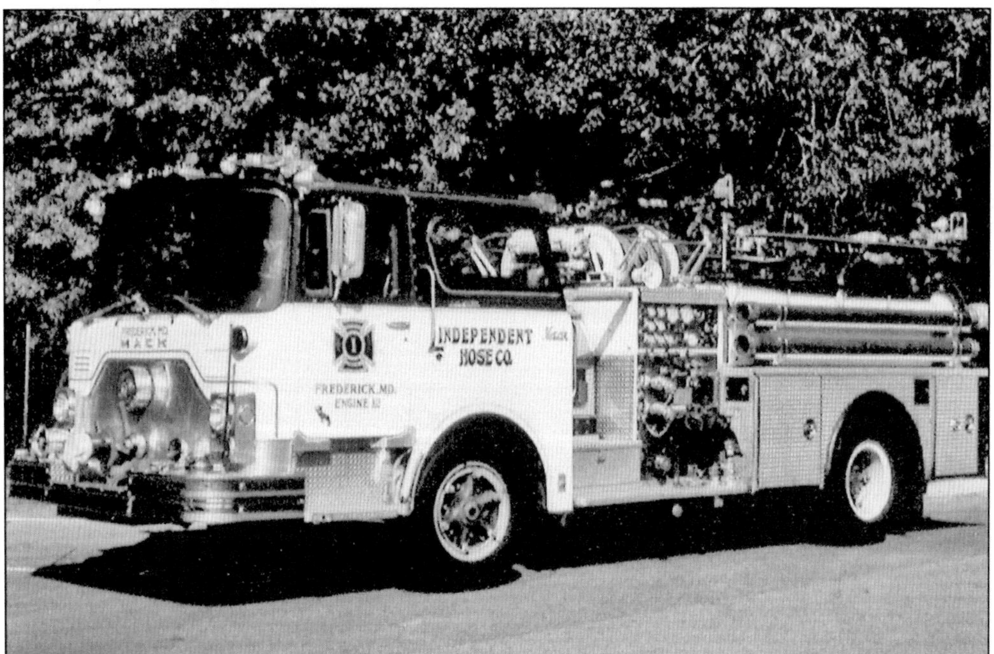

The 1972 CF Mack responded as Engine 12 and was used as the "county engine" before the other city companies were allowed to respond outside of the municipal limits. (Courtesy Warren Jenkins.)

The Firehouse Financial Center is shown here as it looks today with the glass etching. The Independent Hose Company relocated to 310 Baughman's Lane in 1978. (Courtesy Warren Jenkins.)

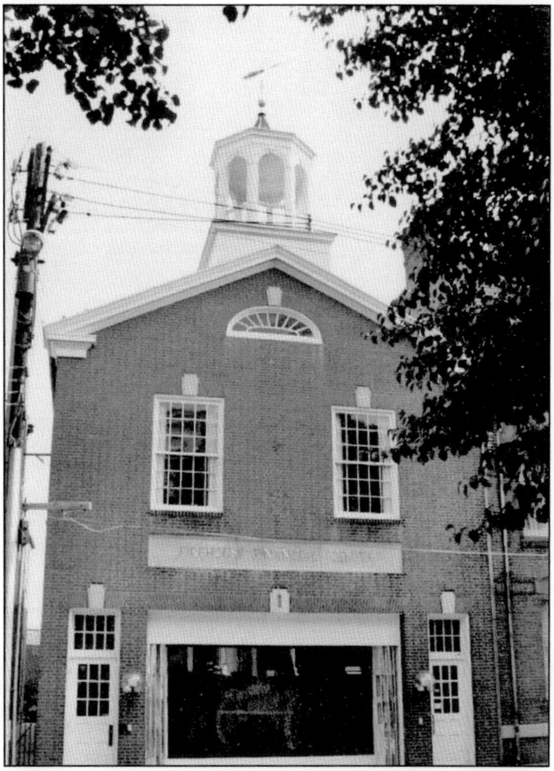

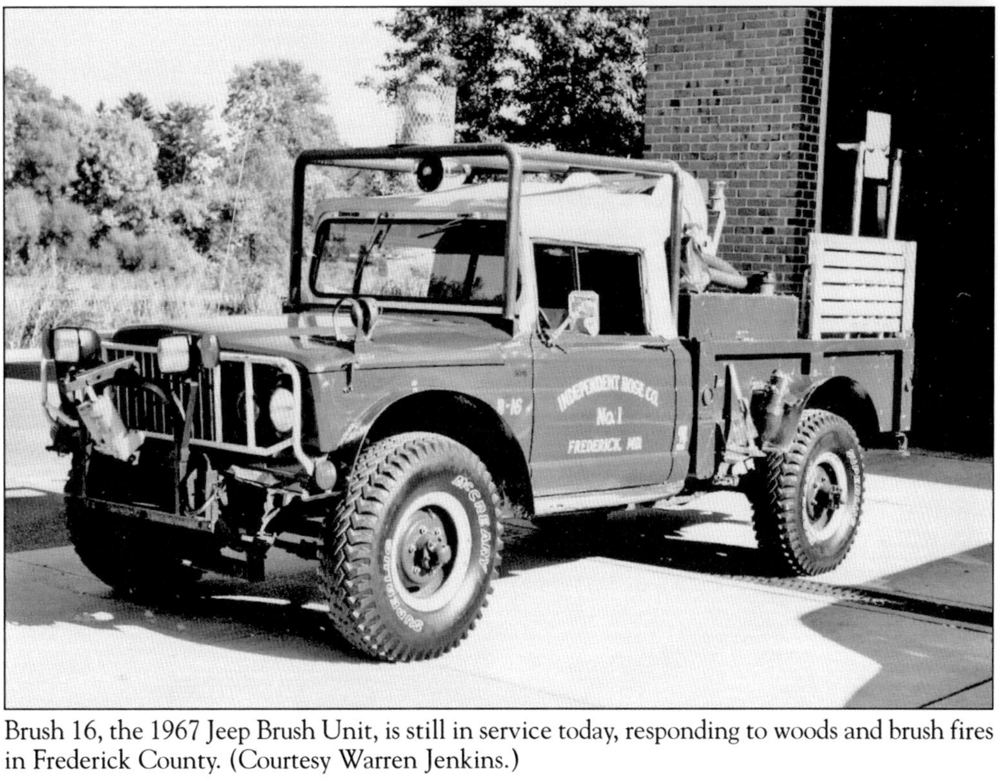

Brush 16, the 1967 Jeep Brush Unit, is still in service today, responding to woods and brush fires in Frederick County. (Courtesy Warren Jenkins.)

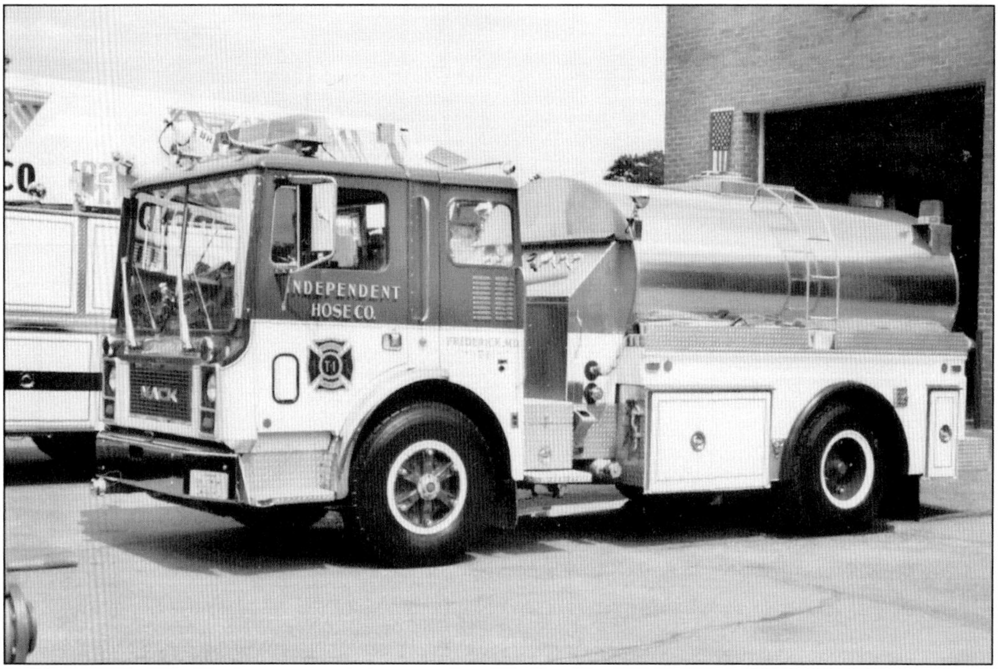

Tanker 1, a 1980 MC Mack 2,000-gallon tanker was purchased by the Independent Hose Company for county water supply response. It was the first MC Mack in service in Maryland. (Courtesy Warren Jenkins.)

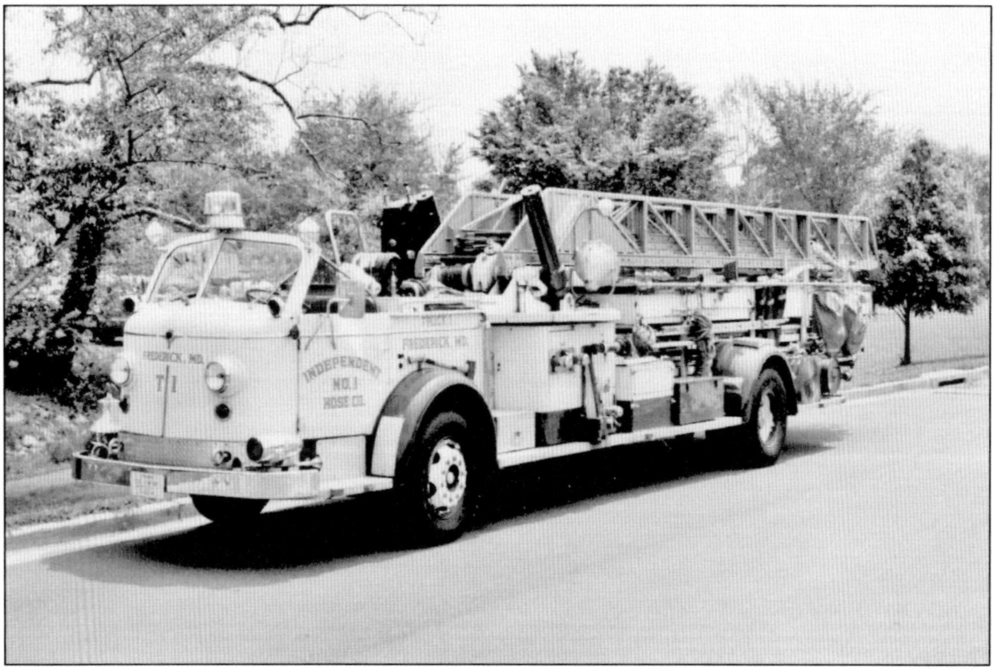

Though ladder service has been available in Frederick City since 1879, the dramatic growth of the area required additional aerial apparatus. In 1982, the Independent Hose Company raised funds to purchase a 1953 American LaFrance aerial ladder that remained in service until being replaced in 1988. (Courtesy Warren Jenkins.)

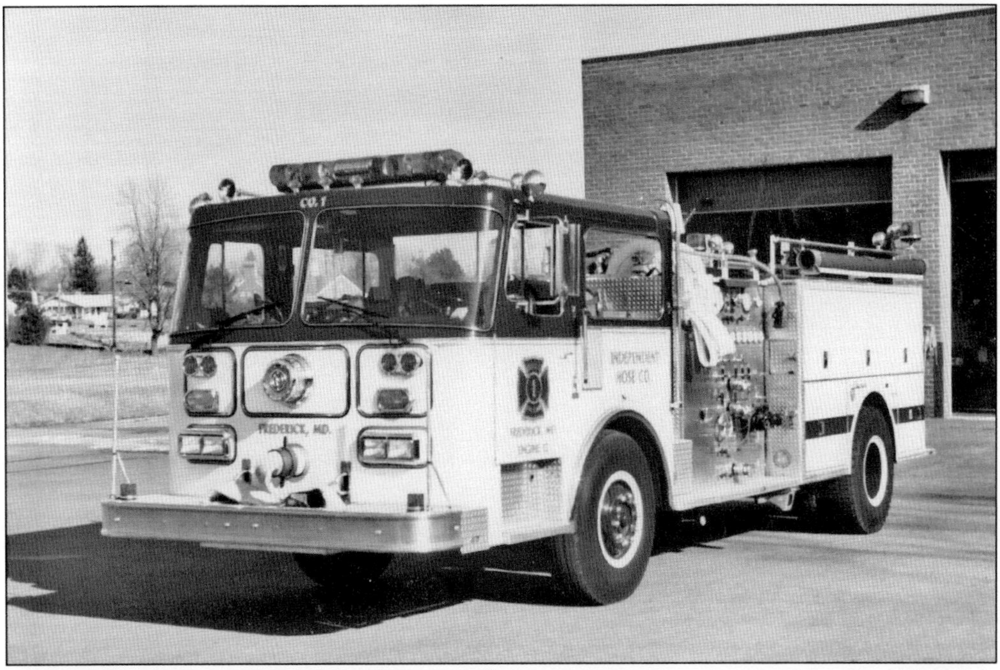

The 1984 Seagrave pumper responded primarily as a county engine until sold to the Hillandale Volunteer Fire Department in Montgomery County, Maryland. (Courtesy Warren Jenkins.)

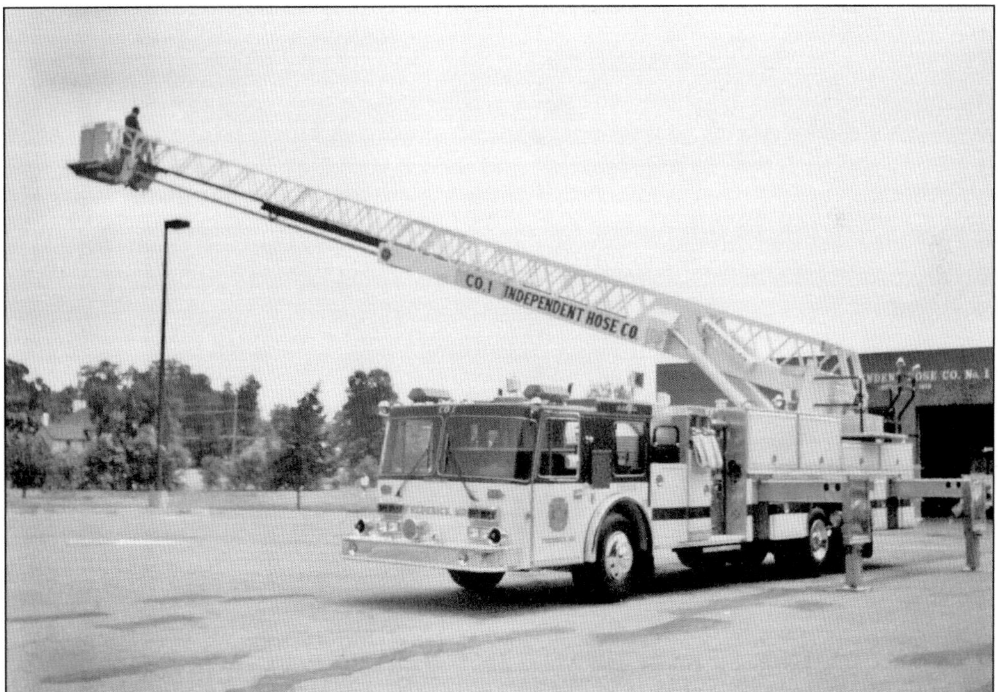

The 1988 Grumman Tower 1 was the first elevated platform unit in Frederick County. The unit, with a 102-foot reach, was purchased with volunteer funds for $420,000. (Courtesy Warren Jenkins.)

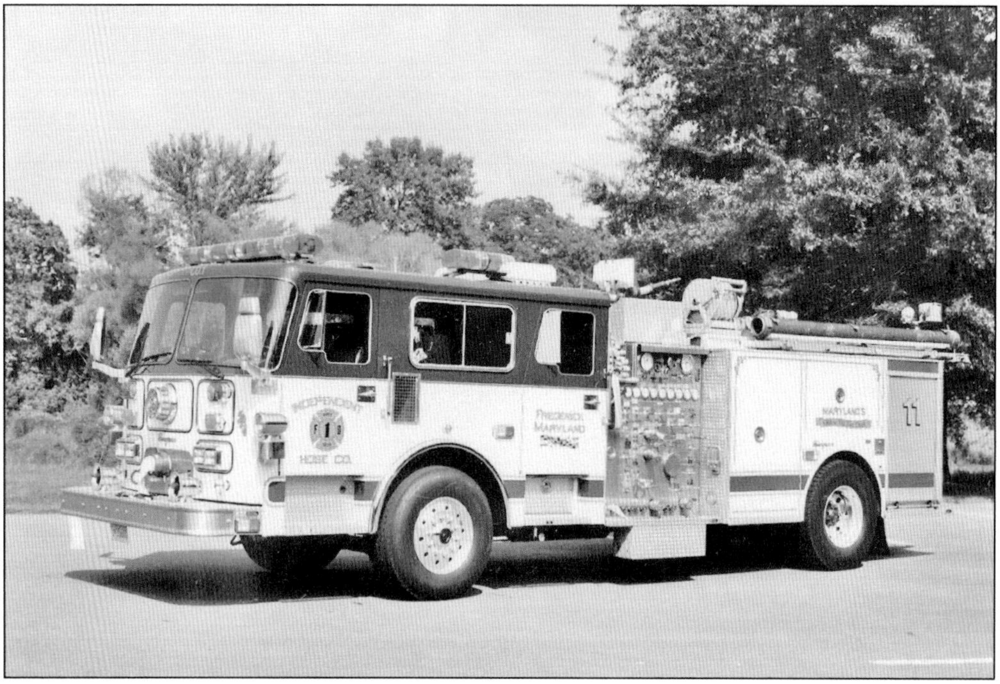

As demands for service increased, the Independent Hose Company rebuilt the 1988 Seagrave to rescue engine standards to allow multiple capabilities, both firefighting and auto extraction. (Courtesy Warren Jenkins.)

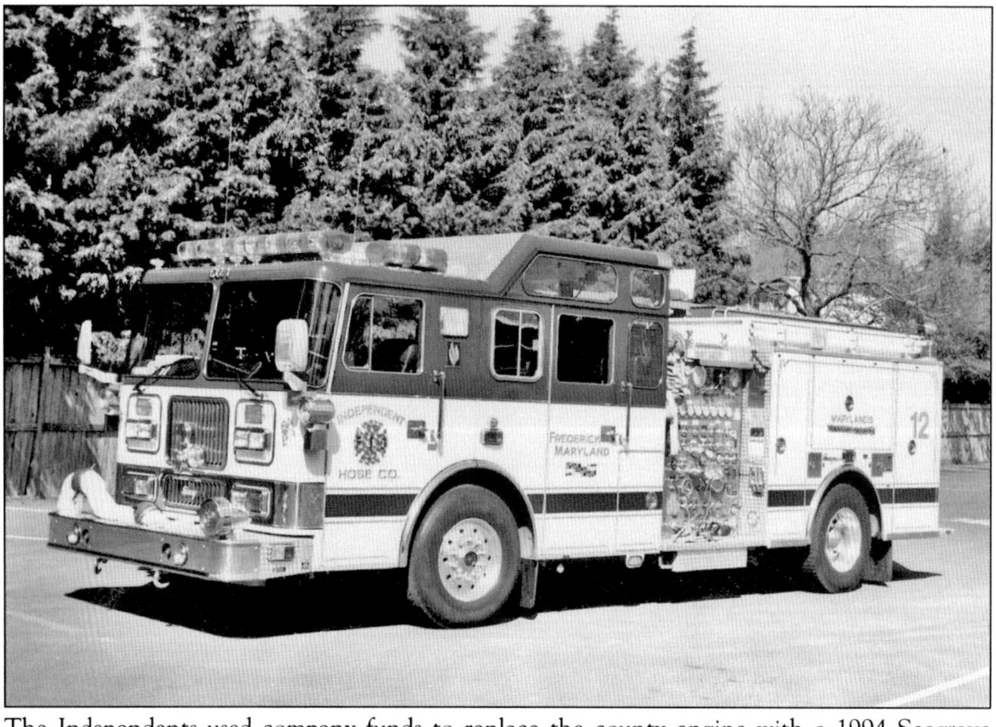

The Independents used company funds to replace the county engine with a 1994 Seagrave Marauder equipped with a 1,500-gpm pump. (Courtesy Warren Jenkins.)

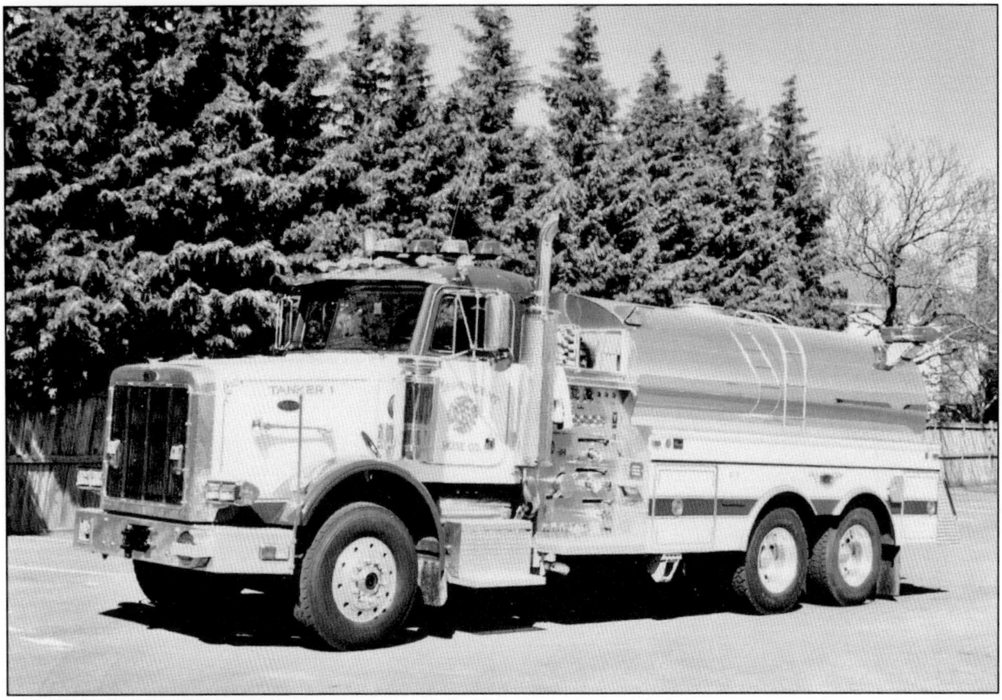

Members of the Independent Hose Company again worked hard to replace the tanker with a 1994 Peterbilt with a tank capacity of 3,000 gallons.

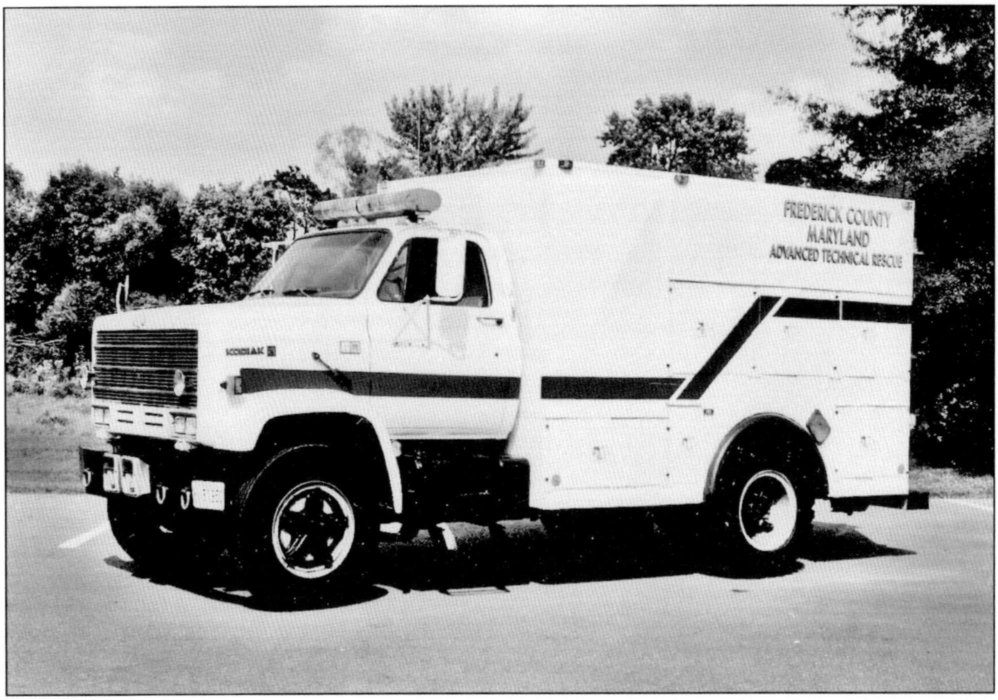

The Independents formed an advanced team to provide specialty rescue service for situations on mountains and cliffs. The Chevrolet Kodiak support vehicle was donated by the Frederick Gas Company.

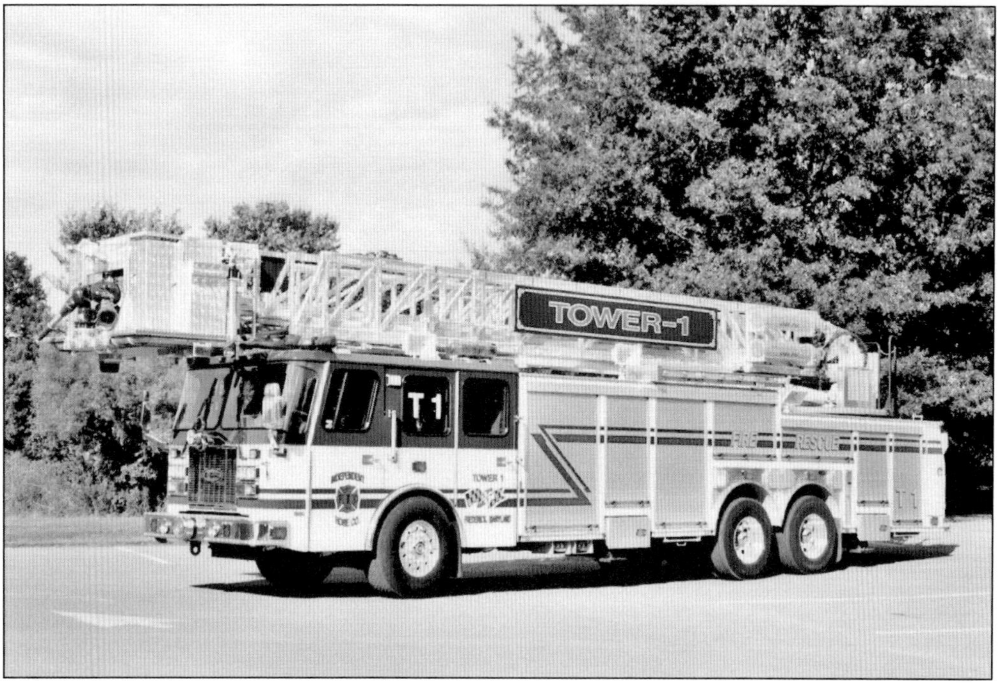

The Independents suffered a tremendous loss when the 1988 ladder tower was destroyed by fire at a maintenance facility in Virginia. It was replaced with a 2001 E-One aerial tower. (Courtesy Warren Jenkins.)

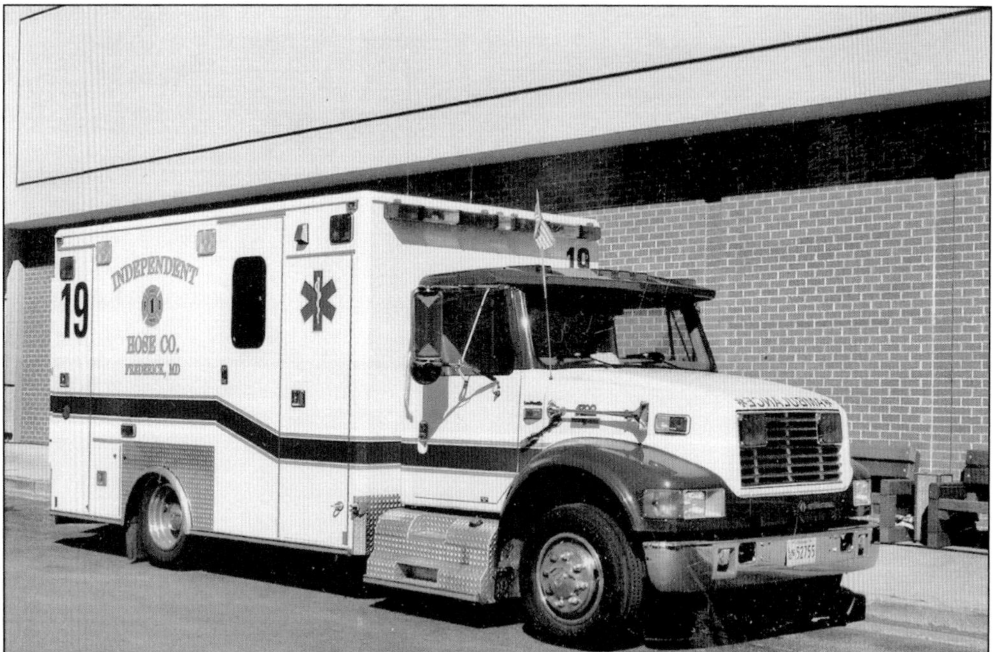

In later years, the City of Frederick started ambulance service out of the Independent Hose Company. This service ultimately became a part of the Independent Hose Company. Shown here is A19, an International Basic Life Support Unit.

Two

JUNIOR FIRE COMPANY NO. 2

The Young Men's Fire Company

In 1838, after a major fire destroyed many houses, several young men of the community met to organize another fire company. They began the fledgling fire company in the quarters of the former Friendship Fire Company. Dubbed the Young Men's Fire Company, they incorporated in 1840 as the Junior Fire Company.

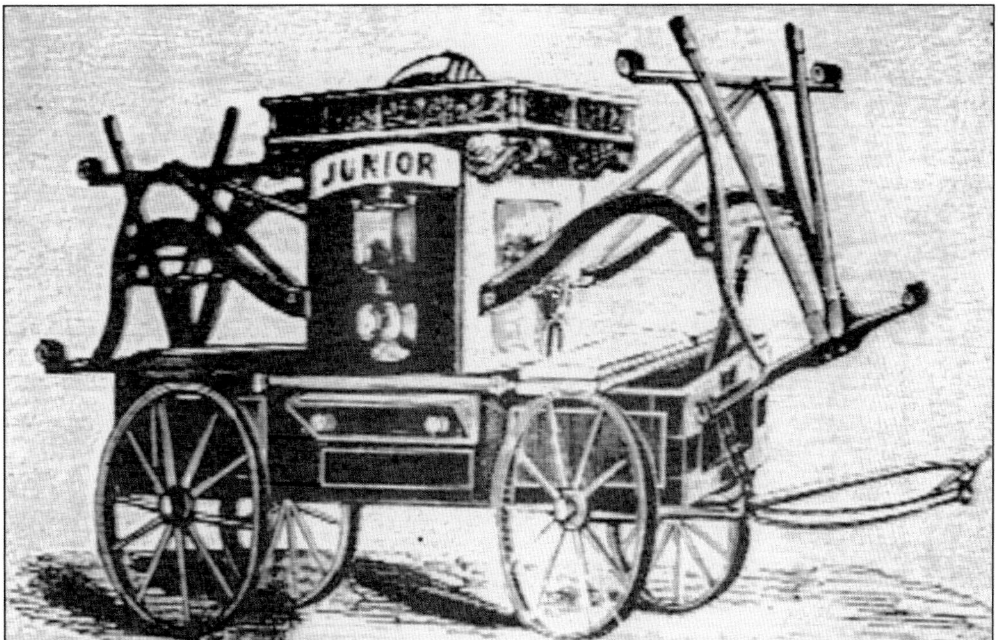

The Junior Fire Company purchased the first fire engine from the Rogers Company of Baltimore in 1839 for the enormous sum of $1,000. The hose was purchased from Dukeharts, also of Baltimore. The pumper was actually the junior model manufactured by Rogers. Thus, the company name was very fitting. (Courtesy JFC [Junior Fire Company No. 2].)

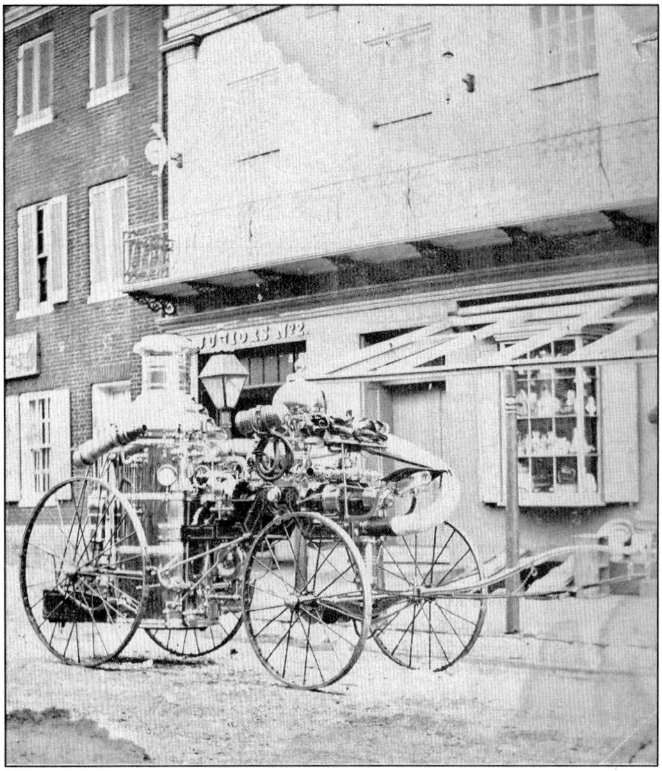

The Juniors purchased the first steam pumper in Frederick in 1876 from the Silsby Manufacturing Company of Seneca Falls, New York. At that time, the words "Steam Fire Engine Company" were added to the official company name. (Courtesy JFC.)

To celebrate the purchase of the engine and the birth of our nation, a massive firemen's parade was held on June 28, 1876. The picture is taken in front of the original engine house at Second and North Market Streets. (Courtesy JFC.)

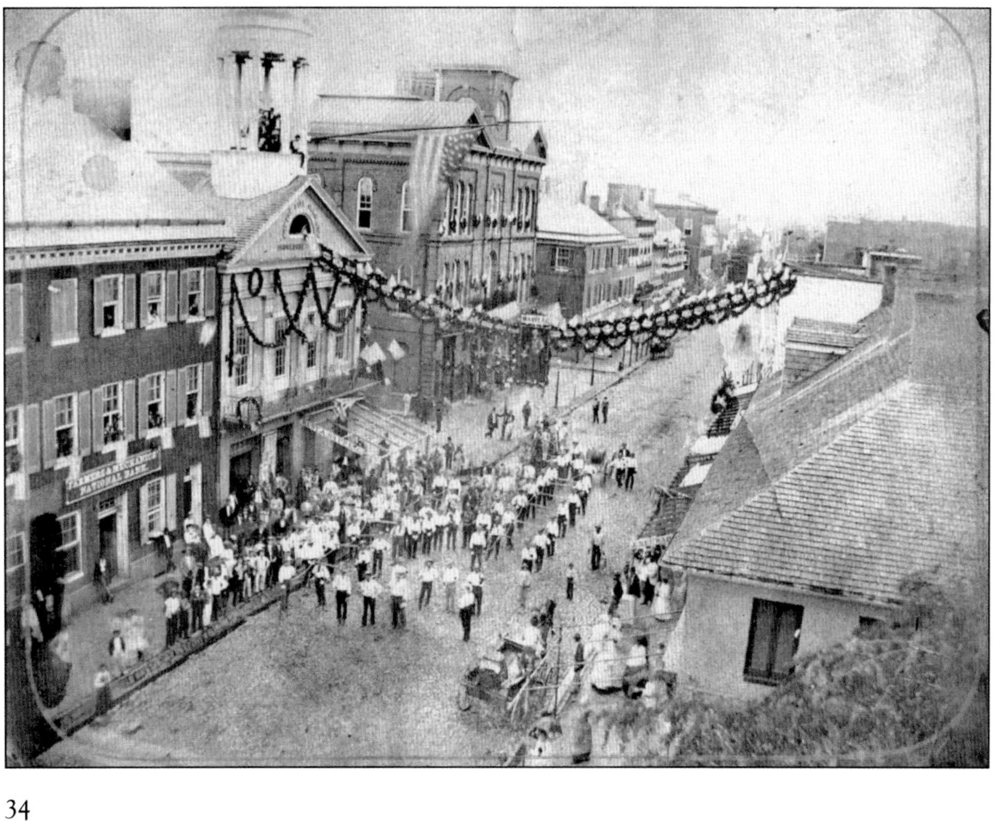

Another parade and celebration occurred in 1889 on the 50th anniversary of the Junior Steam Fire Engine Company No. 2. This picture is believed to show the events. At least two city steamers are shown. (Courtesy JFC.)

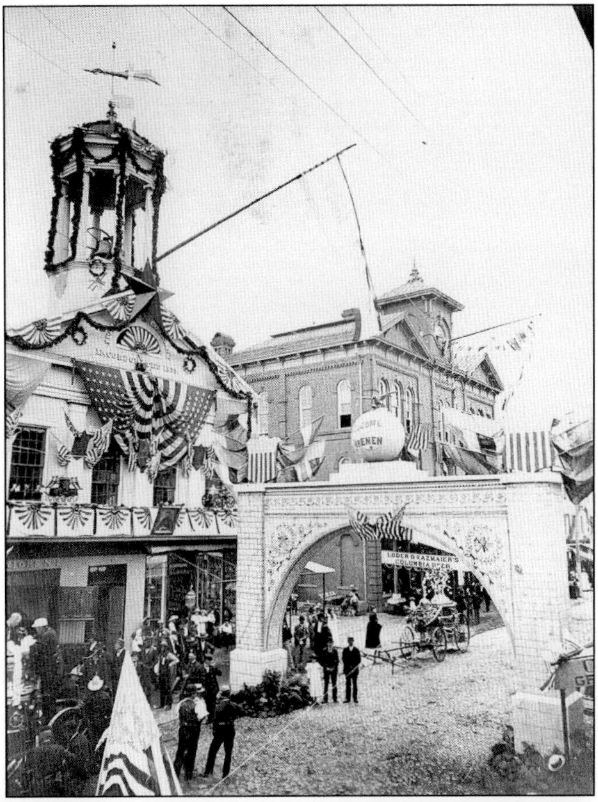

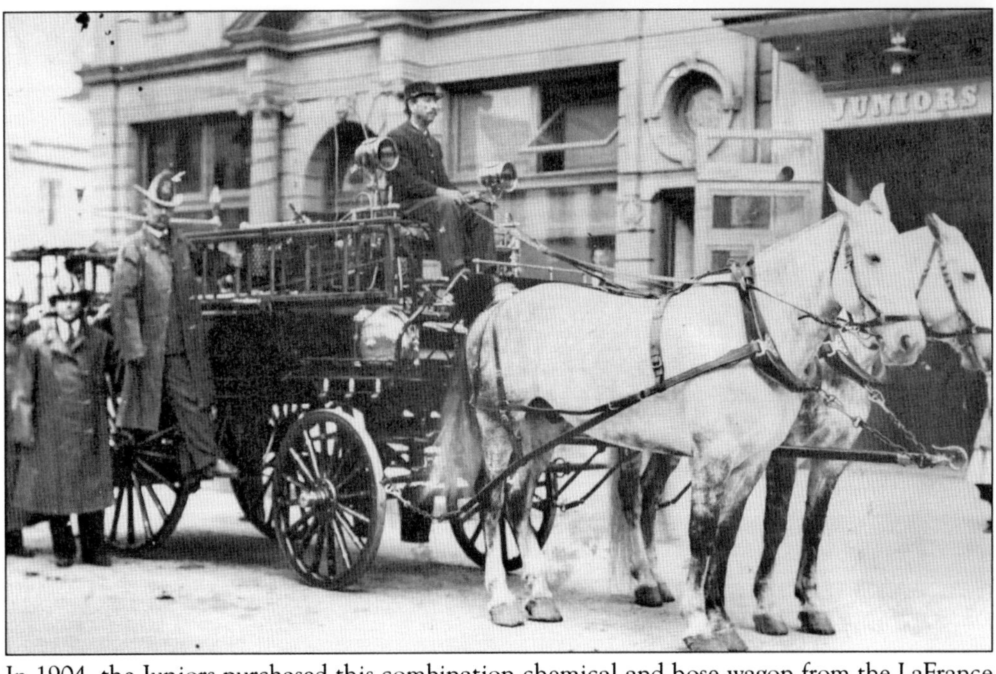

In 1904, the Juniors purchased this combination chemical and hose wagon from the LaFrance Fire Engine Company in Elmira, New York. It replaced an 1889 hand reel and hose carriage. (Courtesy JFC.)

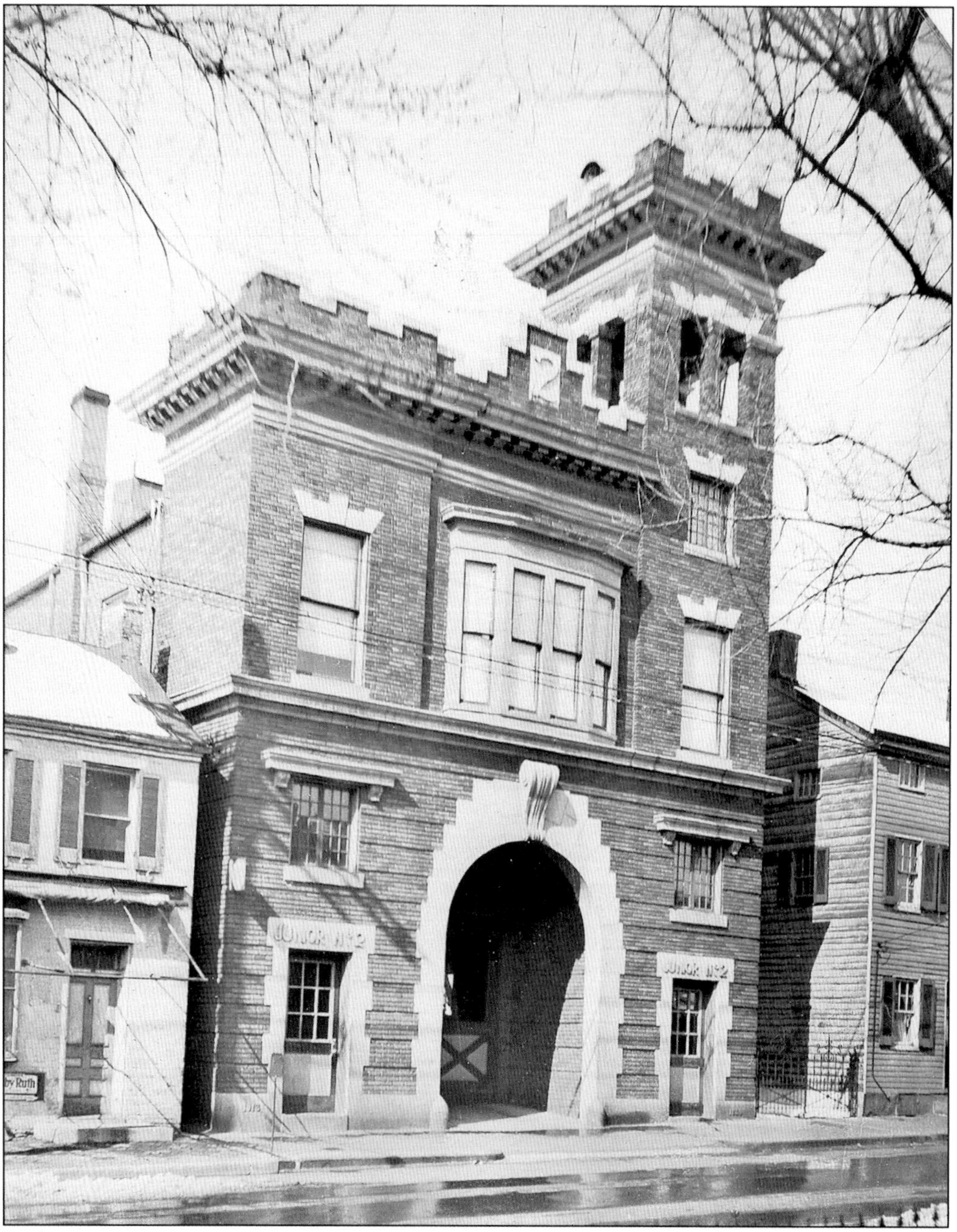

Built in 1913, the Junior Fire Company built the present facility to accommodate horses. However, because of the death of William Davis, horses never occupied this building. Both engines and a ladder truck later responded from this building. (Courtesy JFC.)

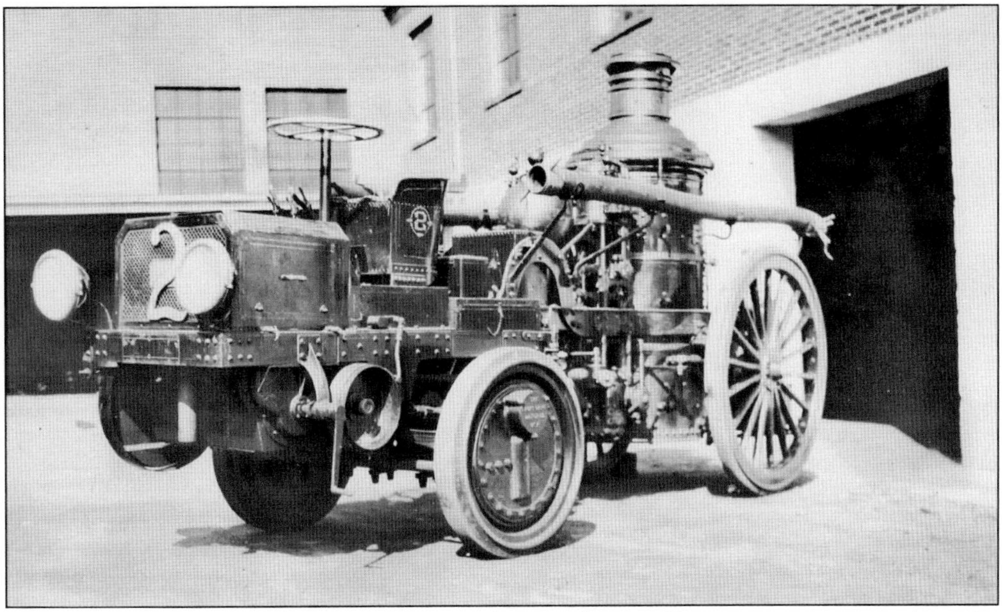

As a result of Davis's death, the Juniors voted to purchase a Christie tractor to attach to the front of the 1908 LaFrance steam fire engine. When the Juniors moved into the new building, the Christie backed the steamer into the engine bay, thus ending the era of horses at the Juniors that began in 1903. (Courtesy UFC [United Steam Fire Engine Company No. 3].)

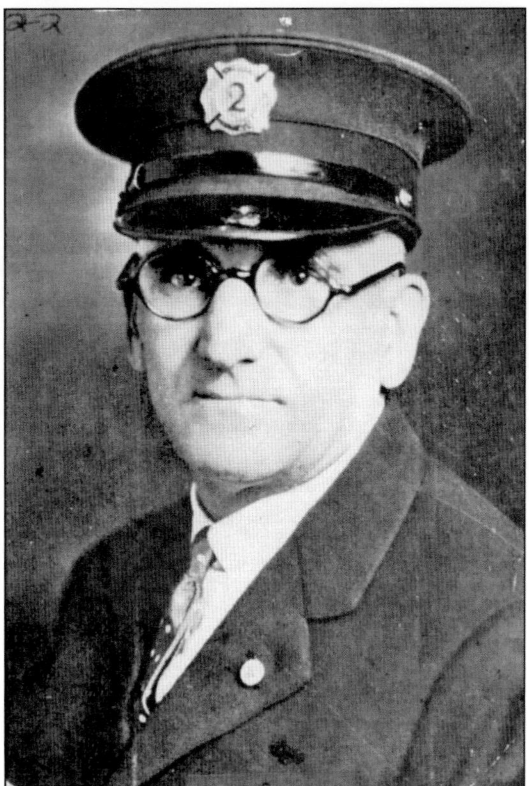

A keystone of the early Juniors membership, George William Shipley served as foreman (later years called captain or chief) from 1902 to 1903 and 1905 to 1934. (Courtesy JFC.)

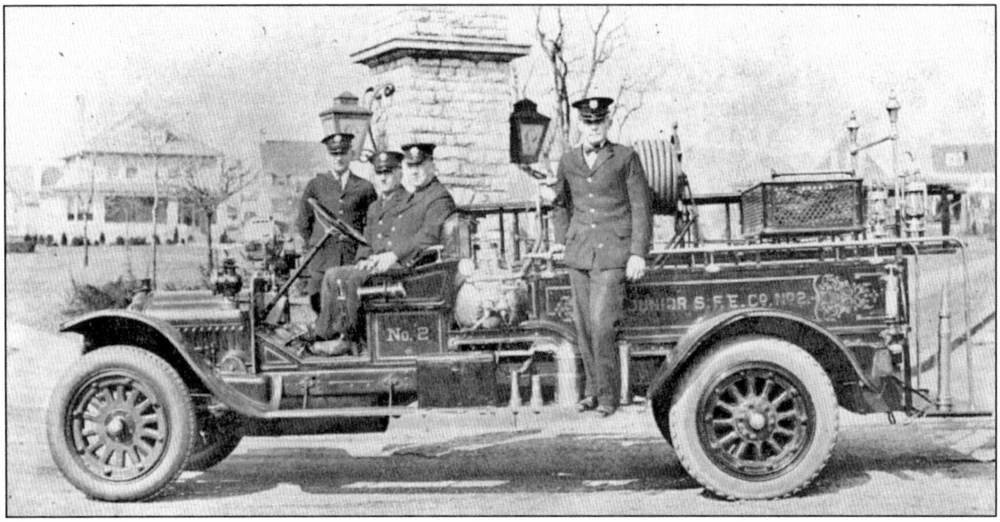

In 1913, a White chassis was delivered and placed under the LaFrance chemical wagon body. This unit responded alone on alarms until the LaFrance returned to service later that year with a Christie tractor. (Courtesy UFC.)

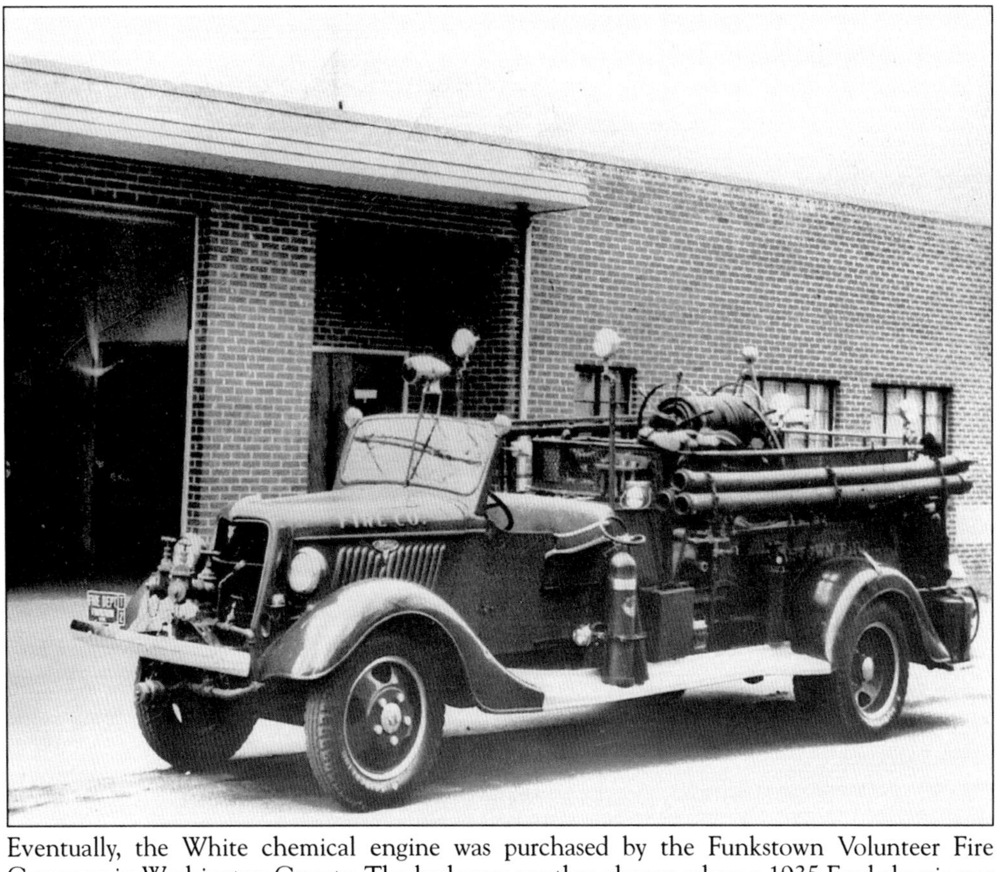

Eventually, the White chemical engine was purchased by the Funkstown Volunteer Fire Company in Washington County. The body saw another change when a 1935 Ford chassis was purchased and placed under the LaFrance body. (Courtesy UFC.)

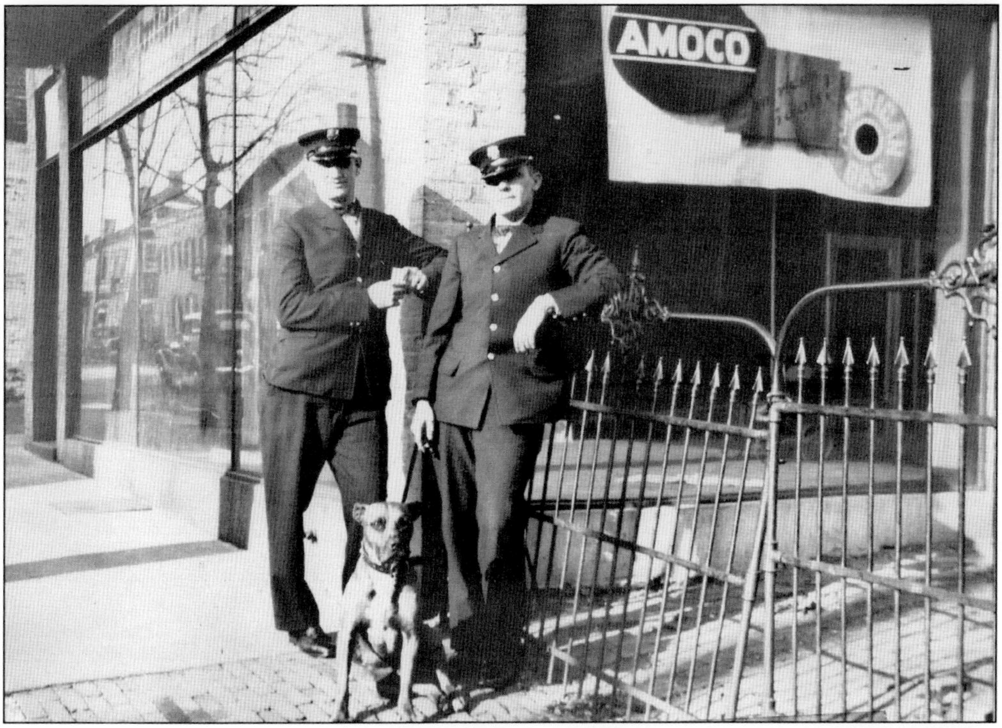

In this 1933 photo, Juniors drivers Ben Markoe and Bill Bopst Sr. pose with their faithful mascot, Fritz. (Courtesy JFC.)

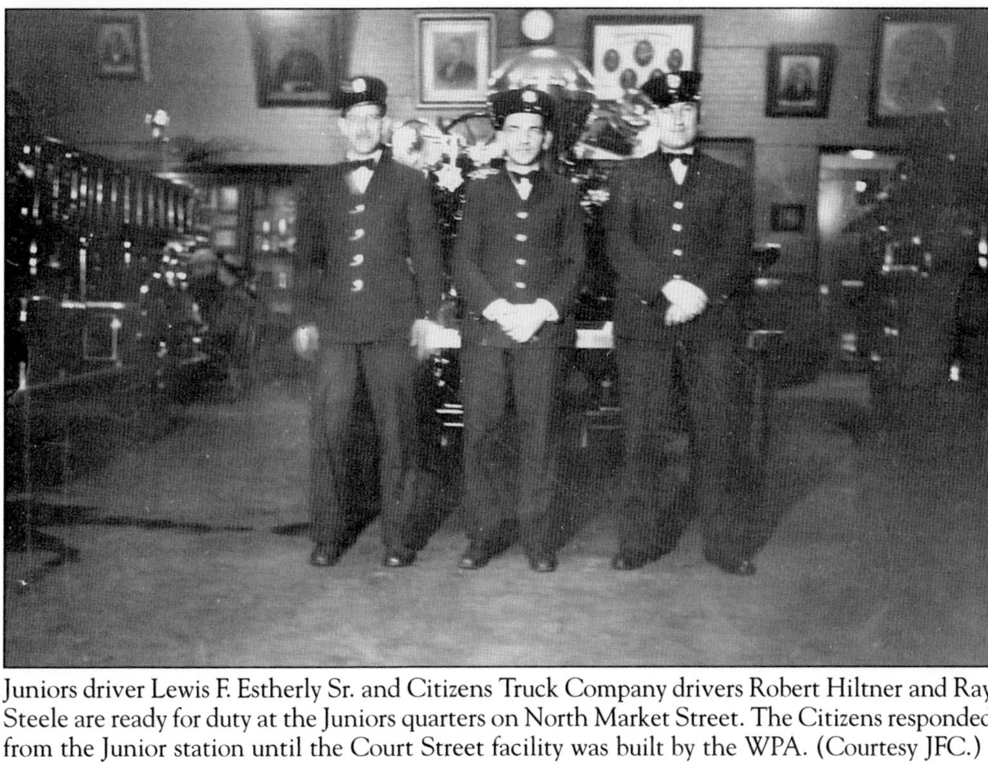

Juniors driver Lewis F. Estherly Sr. and Citizens Truck Company drivers Robert Hiltner and Ray Steele are ready for duty at the Juniors quarters on North Market Street. The Citizens responded from the Junior station until the Court Street facility was built by the WPA. (Courtesy JFC.)

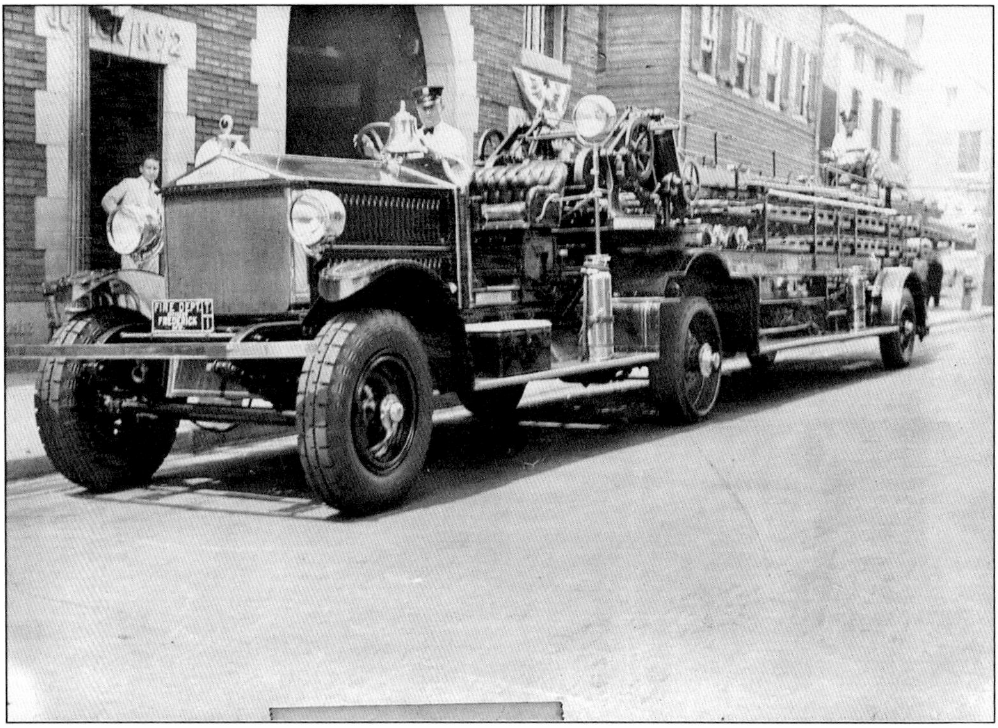

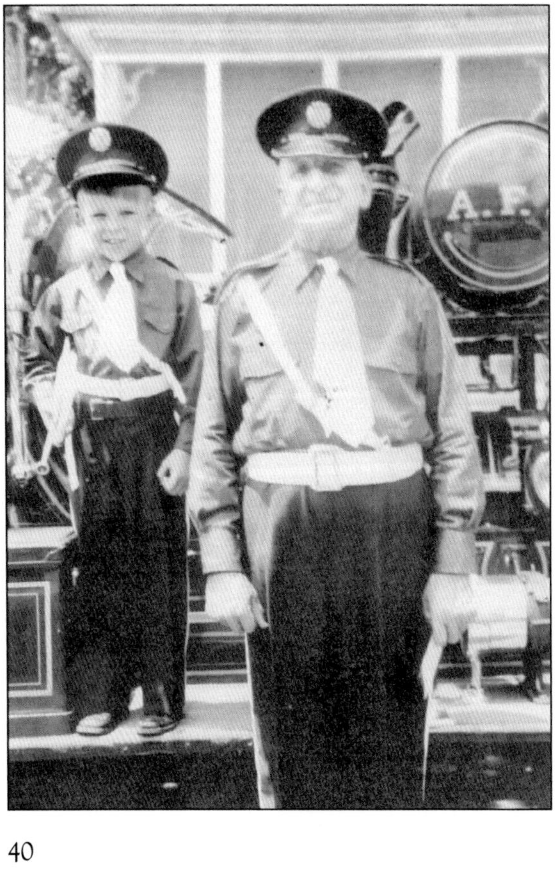

The 1927 Ahrens-Fox ladder truck parked in front of the Junior fire station was actually owned by the Citizens Truck Company. However, the Juniors did purchase the first hand-drawn ladder wagon in Frederick in 1877 from Charles T. Holloway of Baltimore. (Courtesy JFC.)

Tom Fox stands at attention next to his grandfather in this 1936 photo. Tom held many offices in the Juniors and went on to be elected president of the Junior Fire Company No. 2 in 1974. He was ultimately named to the prestigious office of president emeritus. (Courtesy JFC.)

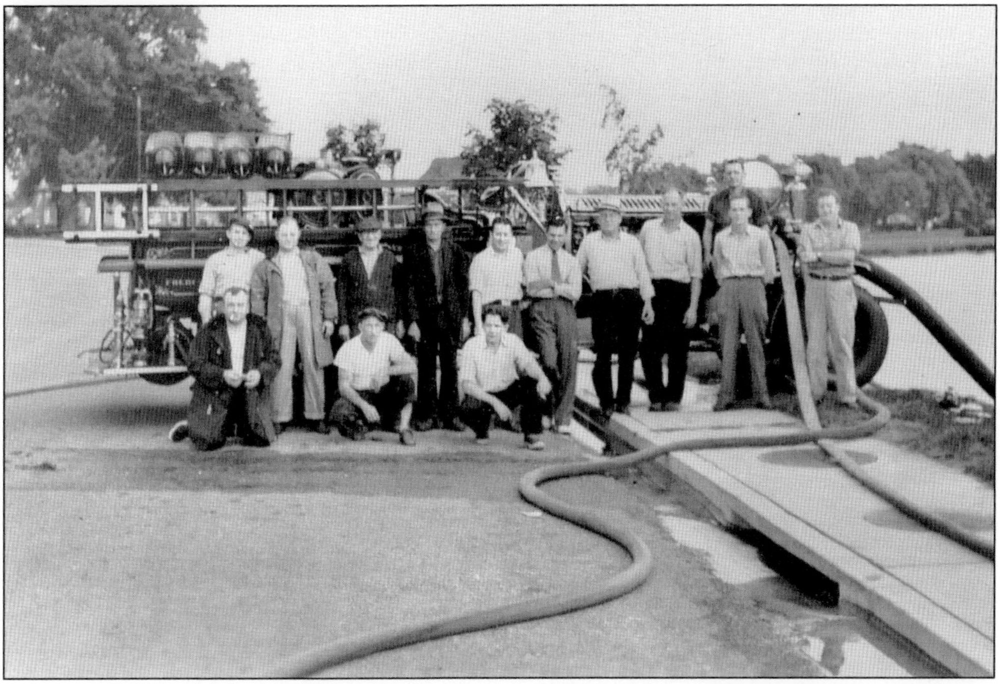

Members of the Juniors pose for a picture next to the 1924 Fox during a pump test at Culler Lake. (Courtesy JFC.)

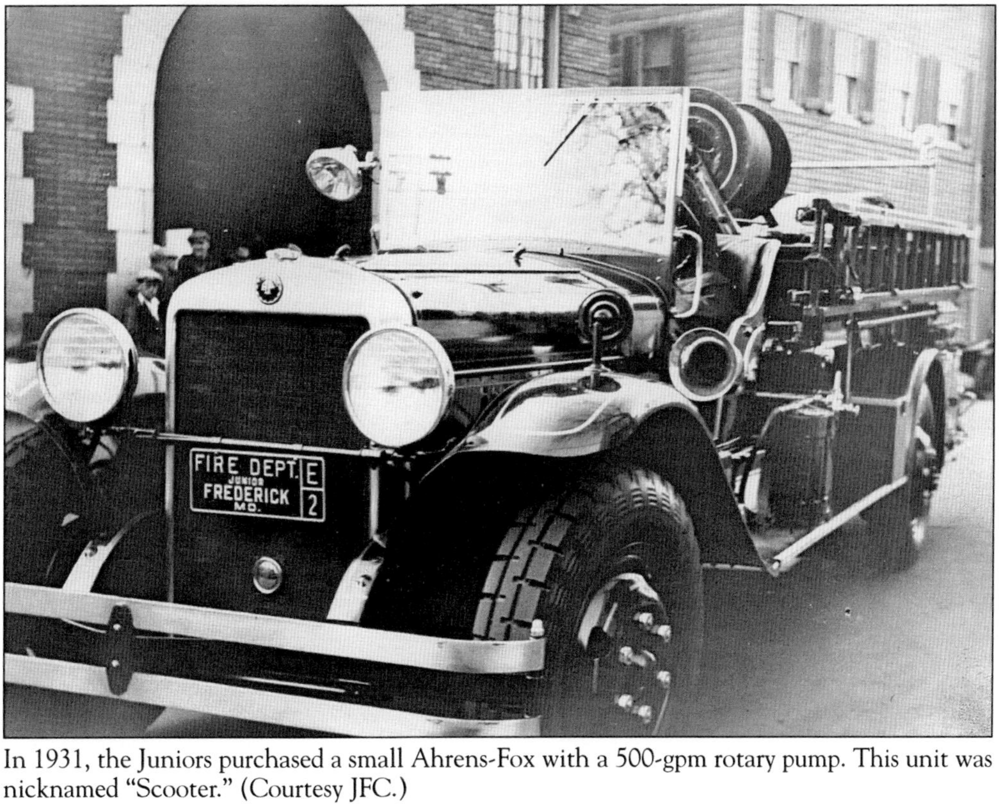

In 1931, the Juniors purchased a small Ahrens-Fox with a 500-gpm rotary pump. This unit was nicknamed "Scooter." (Courtesy JFC.)

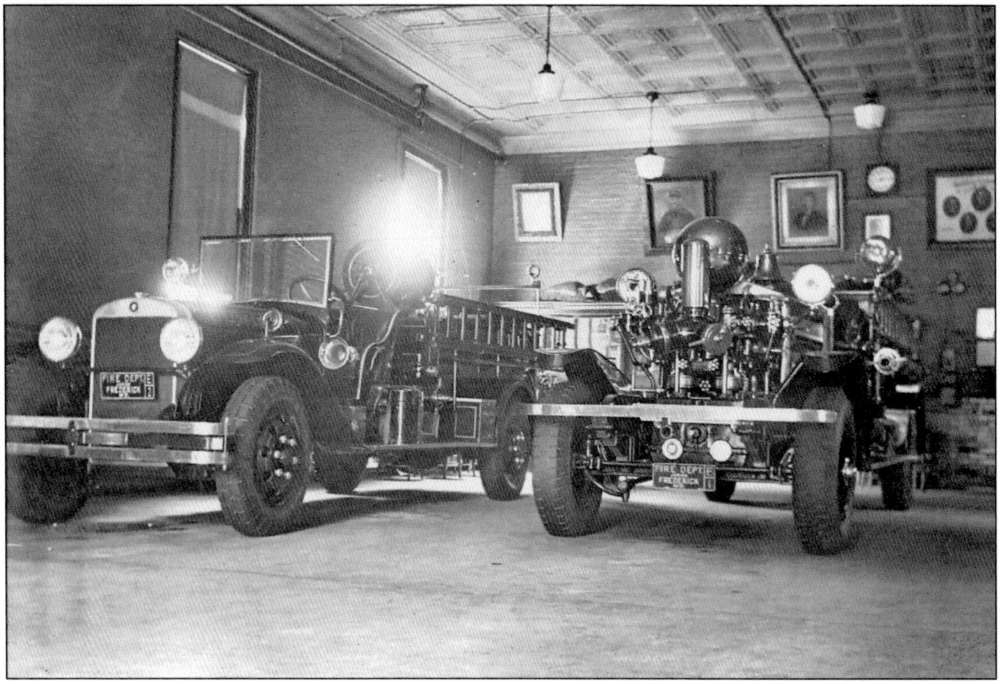

Both of the Ahrens-Fox pumpers owned by the Juniors were housed side by side in the engine room on North Market Street. (Courtesy JFC.)

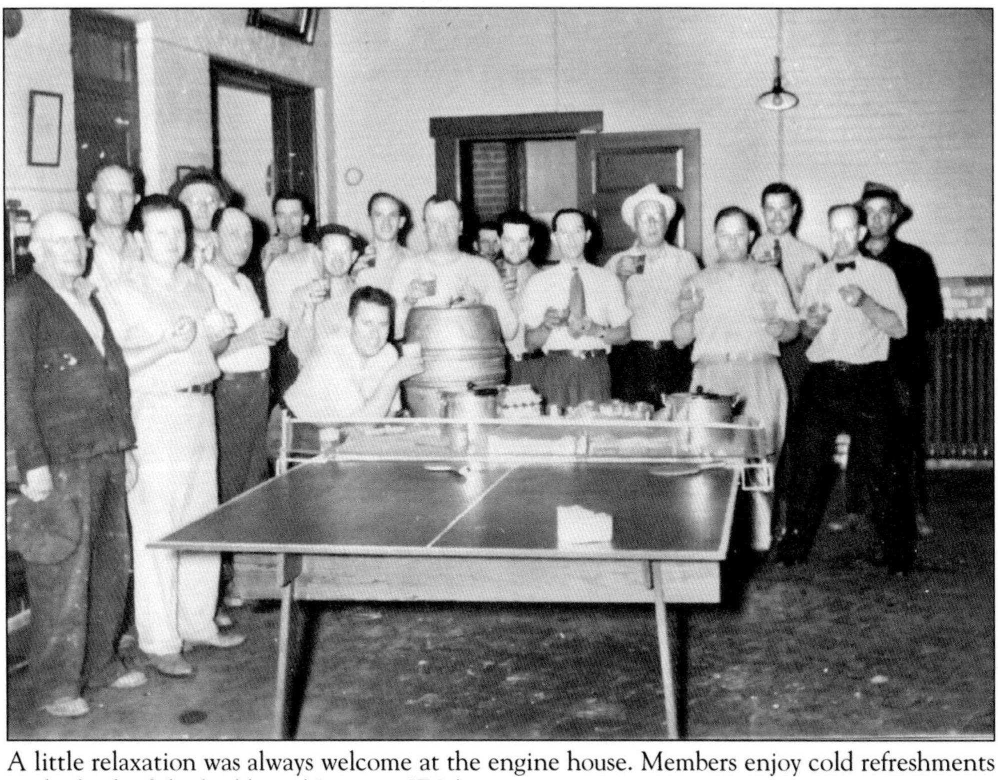

A little relaxation was always welcome at the engine house. Members enjoy cold refreshments in the back of the building. (Courtesy JFC.)

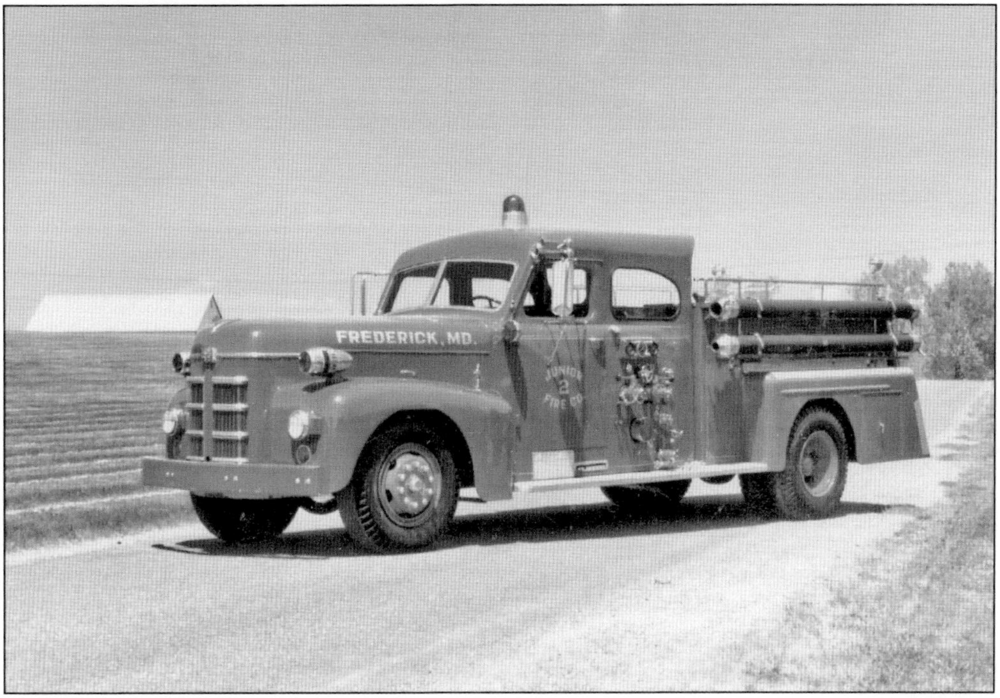

In 1949, the Juniors stayed with tradition by purchasing another Ahrens-Fox with a centrifugal pump. This engine stayed in service until the early 1970s.

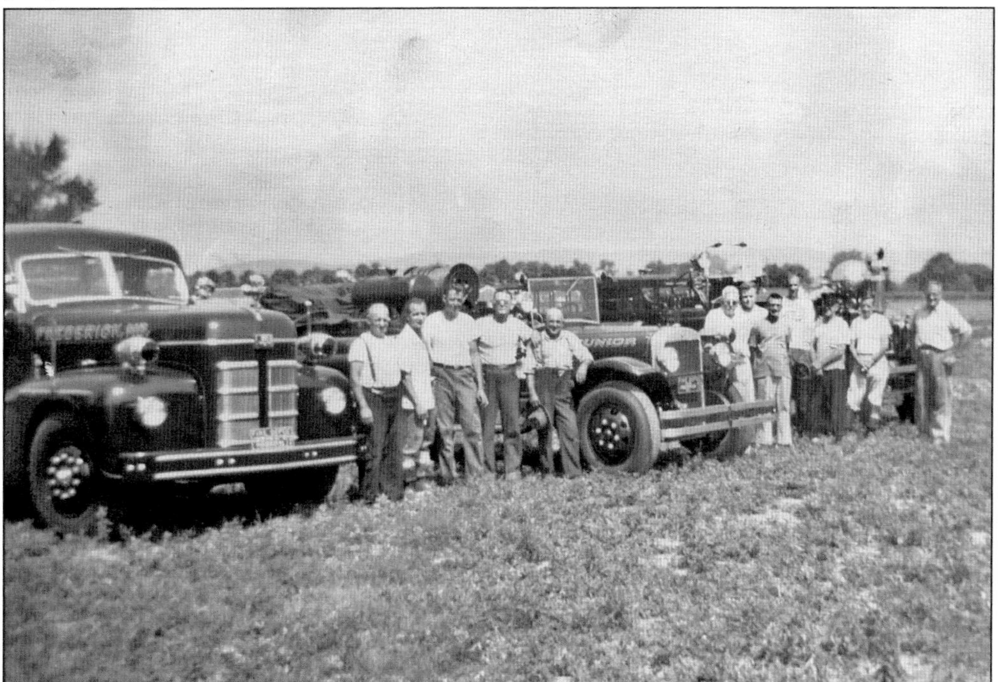

This picture is most likely one of a kind. It is believed the Juniors were the only fire company in the country to have all three types of Ahrens-Fox pumpers in service at the same time. Each engine had a different type of pump. (Courtesy JFC.)

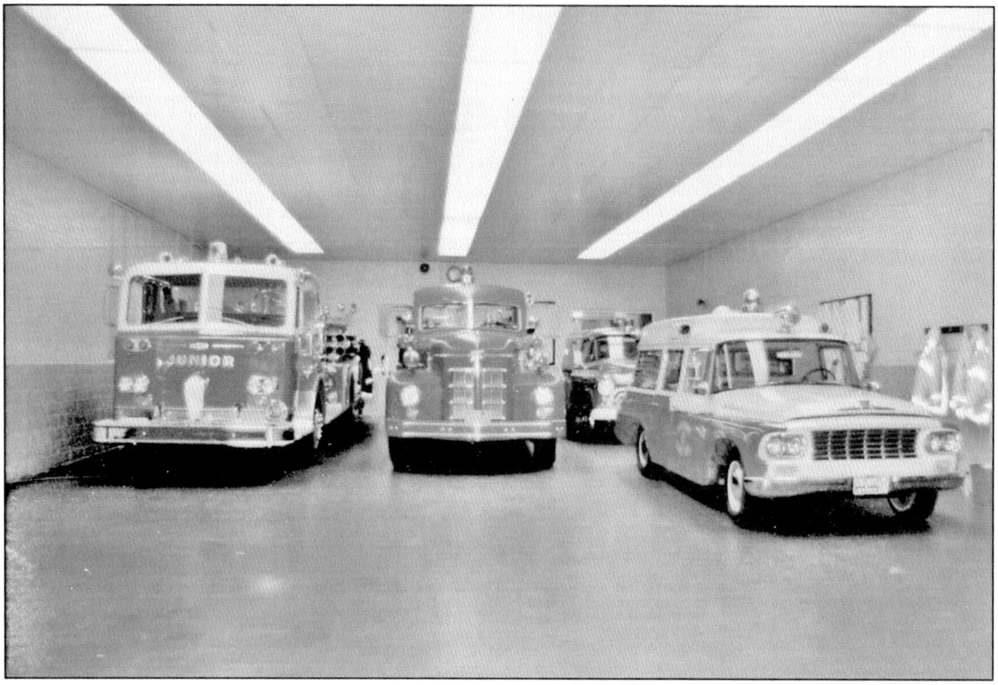

Crowded into the engine room are the 1971 Pirsch, 1949 Ahrens-Fox, 1961 International ambulance, and 1957 Chevrolet utility truck. (Collection of Charles M. Hahn.)

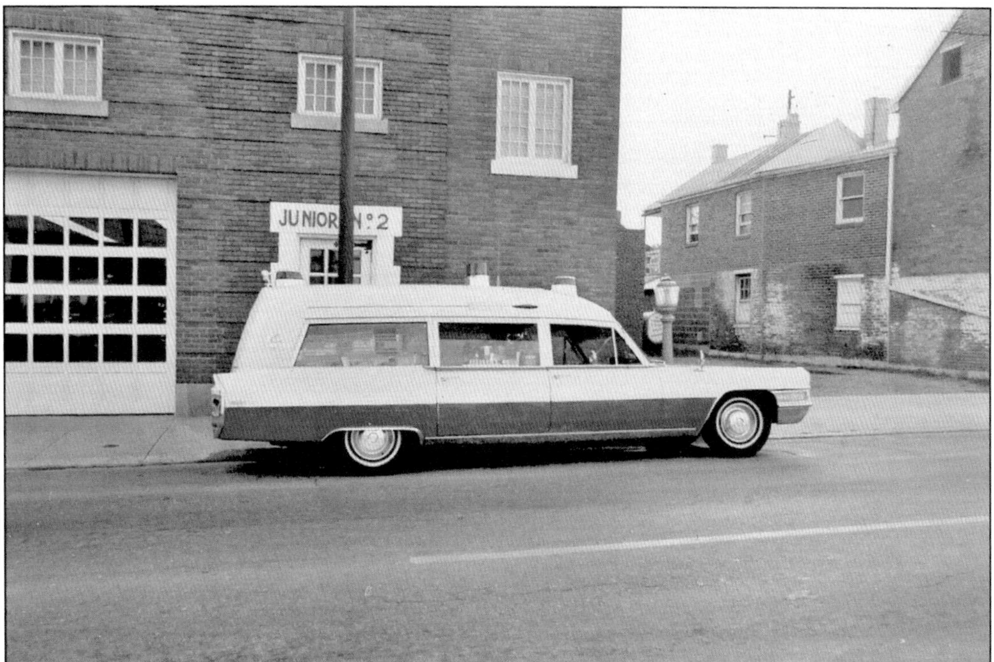

The Juniors entered the ambulance service in September 1965 when they placed this new Cadillac Miller Meteor in service. The volunteer crews were made up of members of the Juniors and Independents and volunteers from Fort Detrick. (Collection of Charles M. Hahn.)

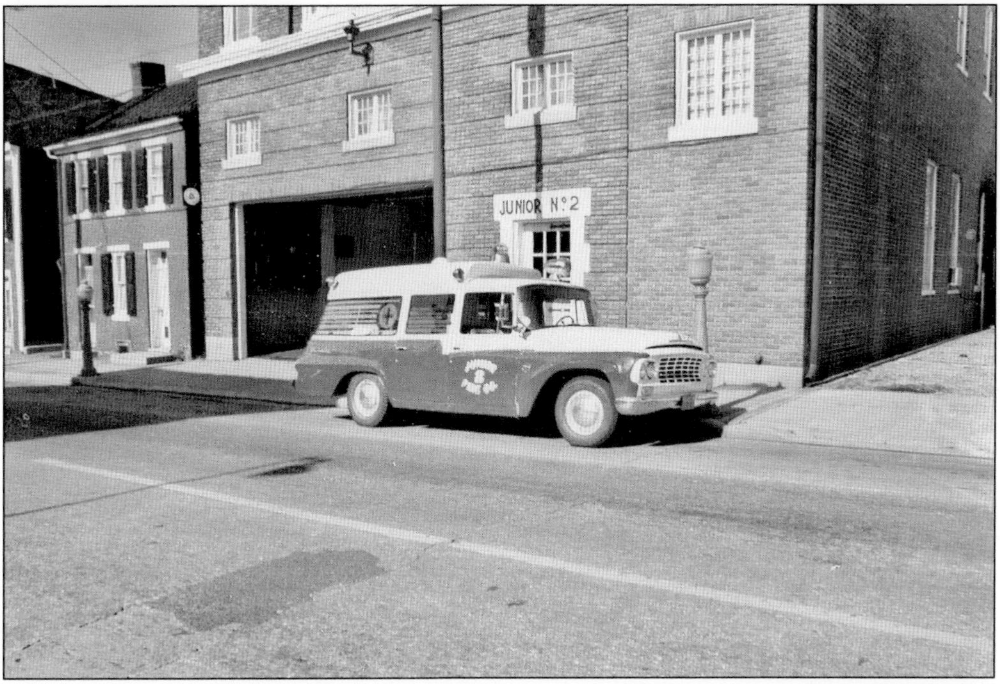

After only two months in service, the Cadillac was severely damaged in an accident with a snow plow during a severe winter storm. The Juniors borrowed and eventually purchased a 1961 International ambulance from the Greenbelt Fire Department in Prince George's County. (Collection of Charles M. Hahn.)

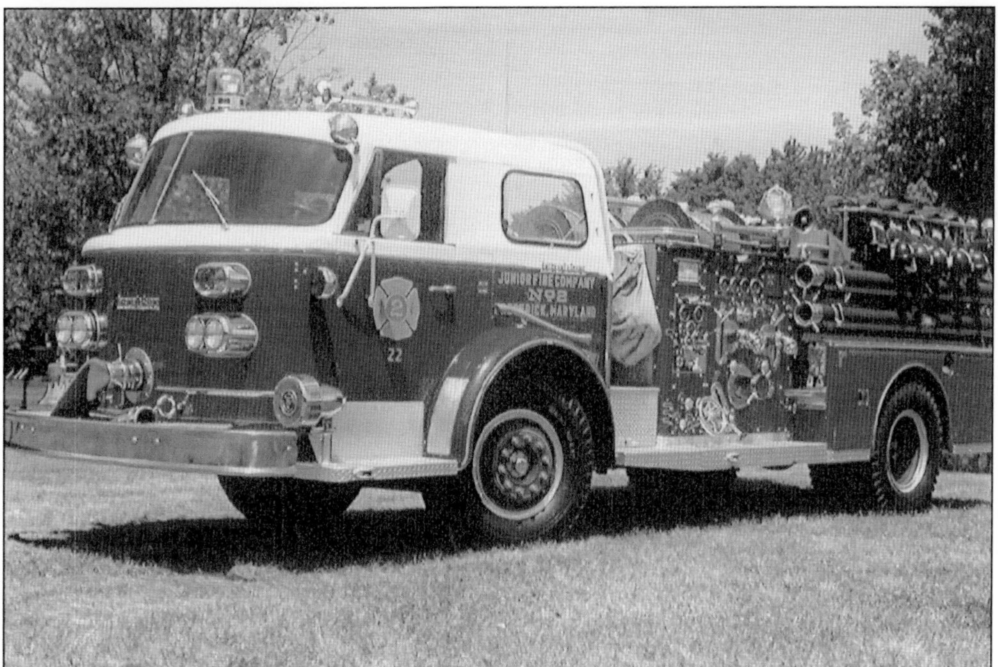

Just as the Juniors favored Ahrens-Fox, they also had an affection for LaFrance, as indicated by the purchase of the 1959 American LaFrance engine. (Courtesy Warren Jenkins.)

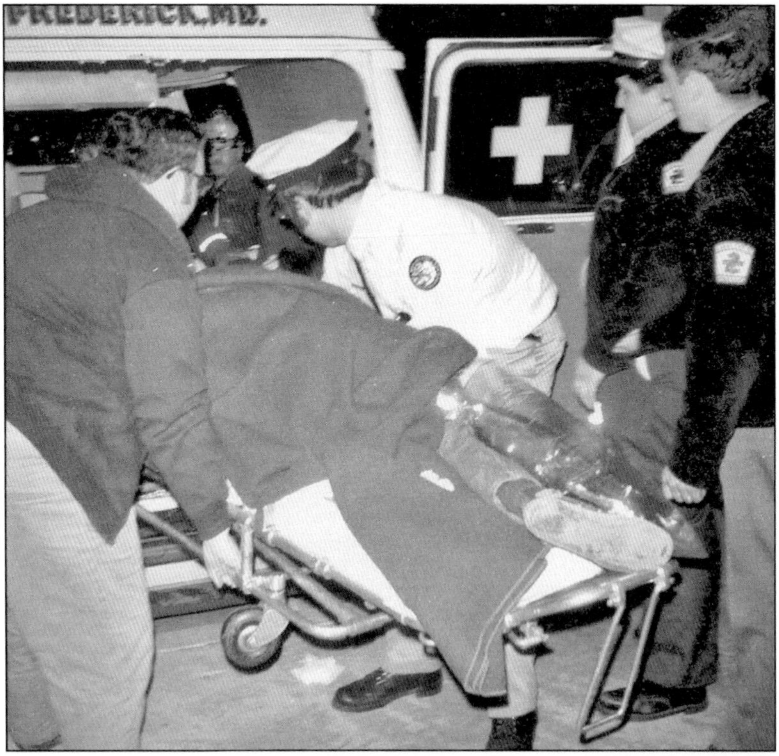

Junior ambulance crew John Knipple (left) and Steve Carter (right) load a patient from an auto accident under the watchful eye of John Hahn and an unidentified Juniors member.

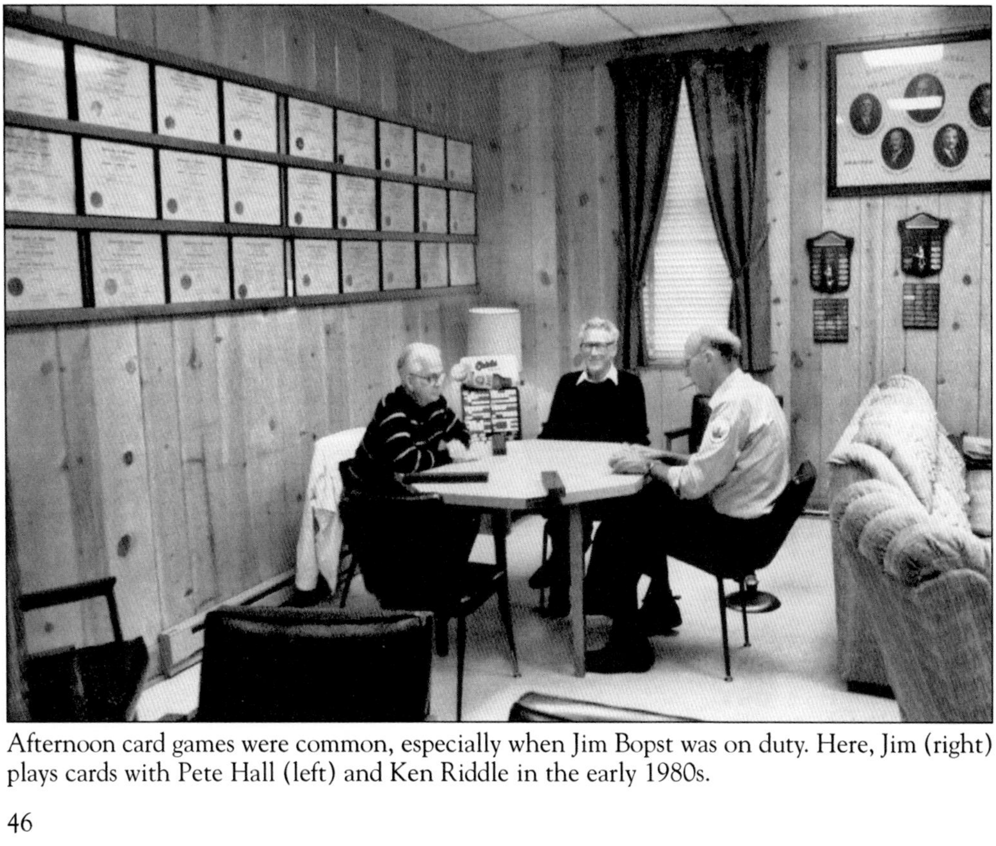

Afternoon card games were common, especially when Jim Bopst was on duty. Here, Jim (right) plays cards with Pete Hall (left) and Ken Riddle in the early 1980s.

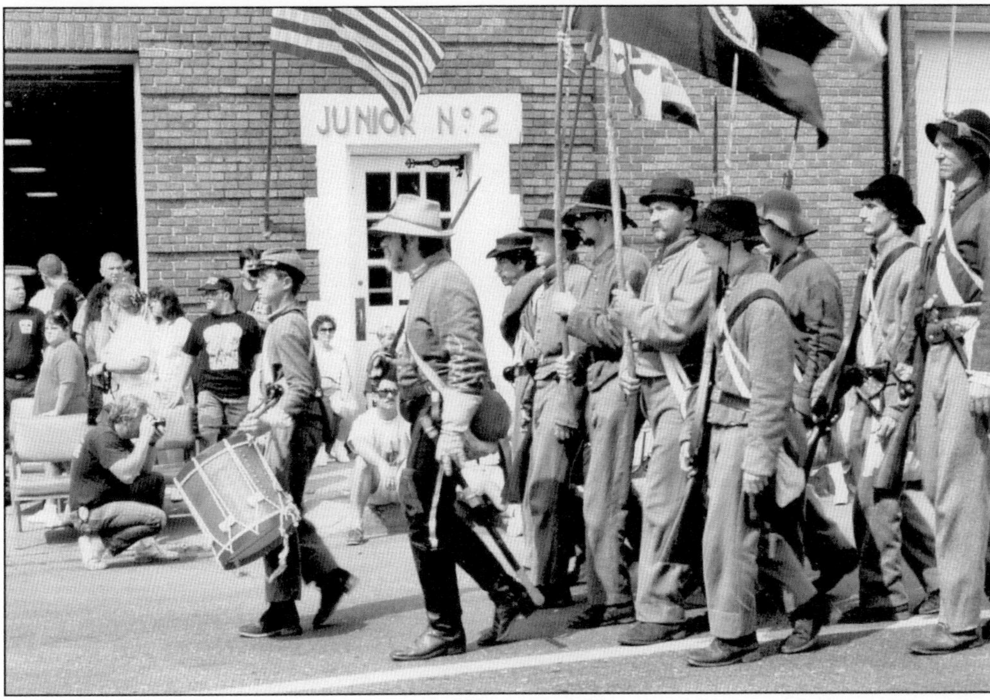

Civil War re-enactors pass the Juniors during the 250th anniversary parade of the city of Frederick. It should be noted the original Junior engine house at Second and North Market Streets was utilized as a part of General Hospital #6 during the Battle of Antietam. Both Union and Confederate troops were treated at the "Juniors Hall." (Courtesy H. Richard Hahn.)

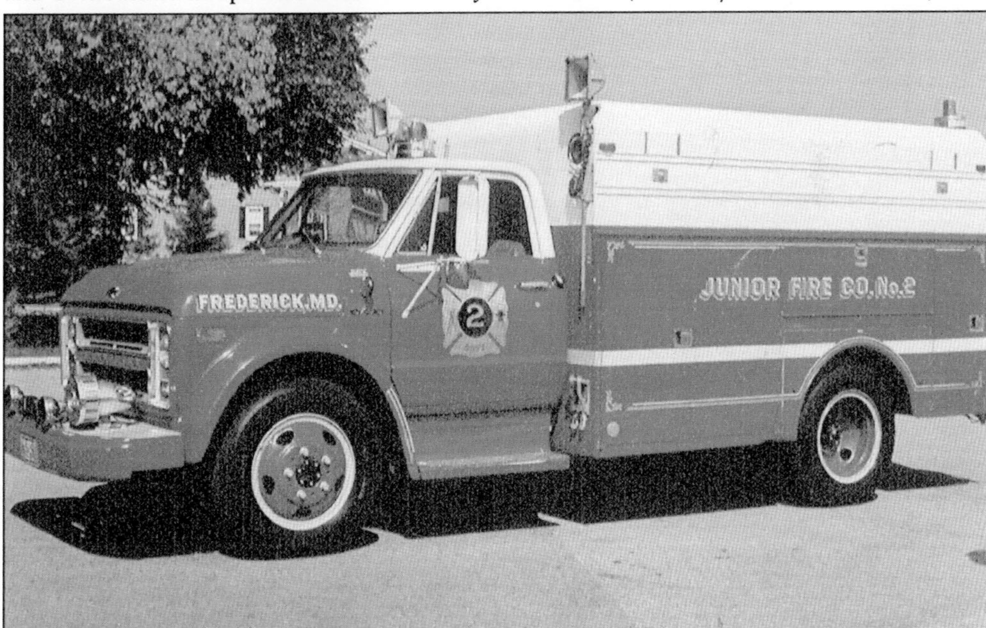

In 1971, the Junior Fire Company expanded services to include a rescue squad. Squad 2 was a 1971 Chevrolet. The 1957 utility truck served as a light duty squad prior to this purchase. (Courtesy Warren Jenkins.)

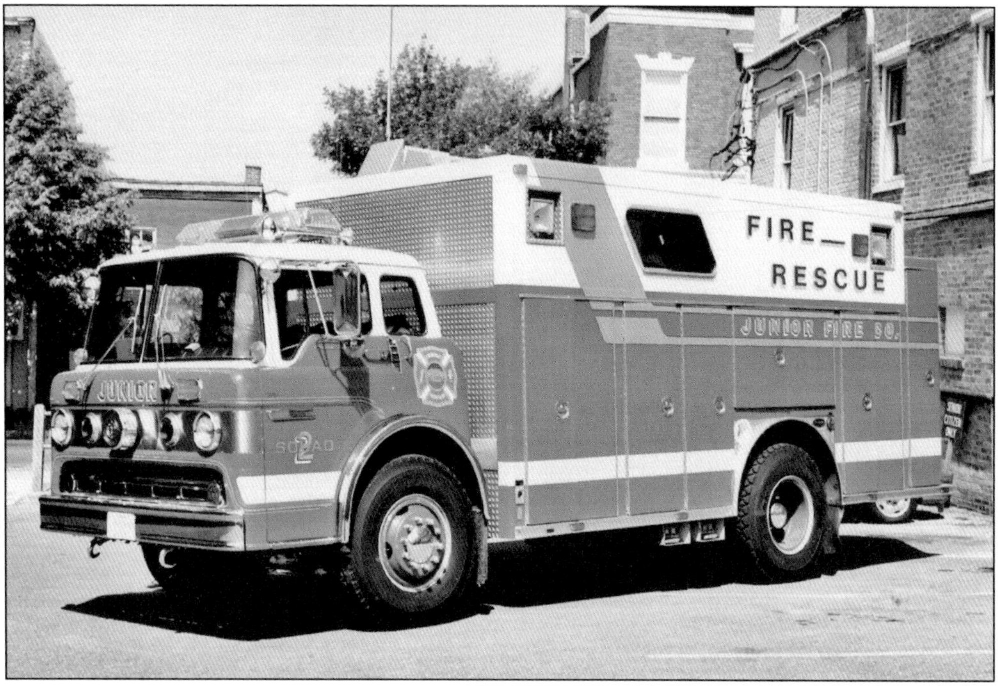

The Juniors continued rescue service with the purchase of the 1987 Ford E-One, replacing the 1971 Chevrolet. The new squad included a four-bottle cascade system. This unit was replaced in August 2004 with a 2004 Pierce. (Courtesy Warren Jenkins.)

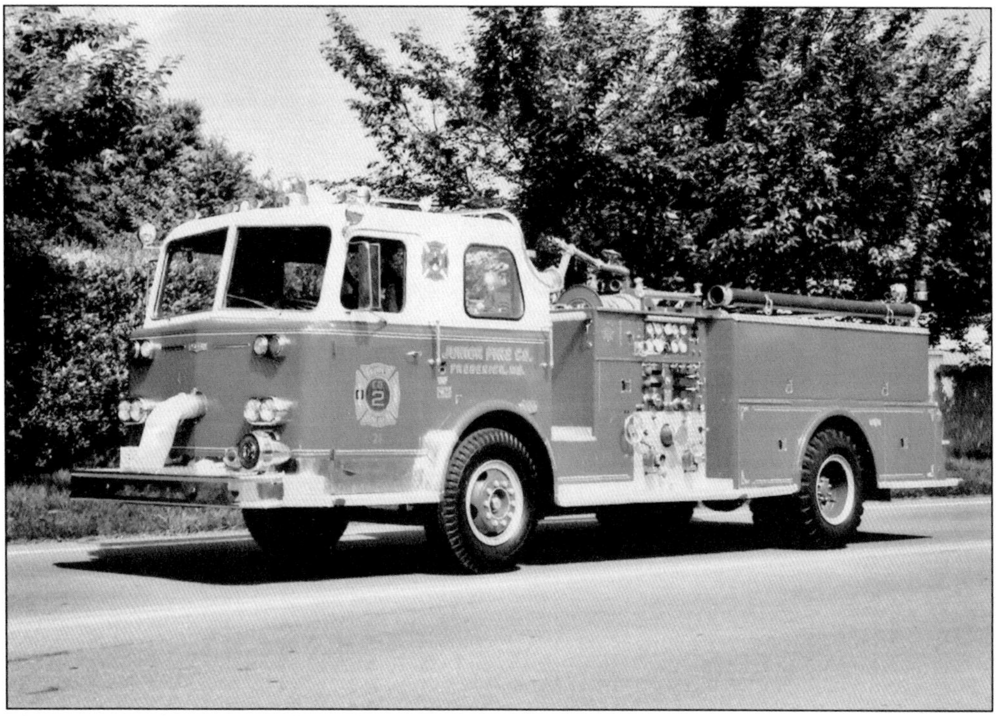

The 1971 Pirsch was recently sold to Steve Rogers of Frederick. He has restored the pumper to nearly original condition. (Courtesy Warren Jenkins.)

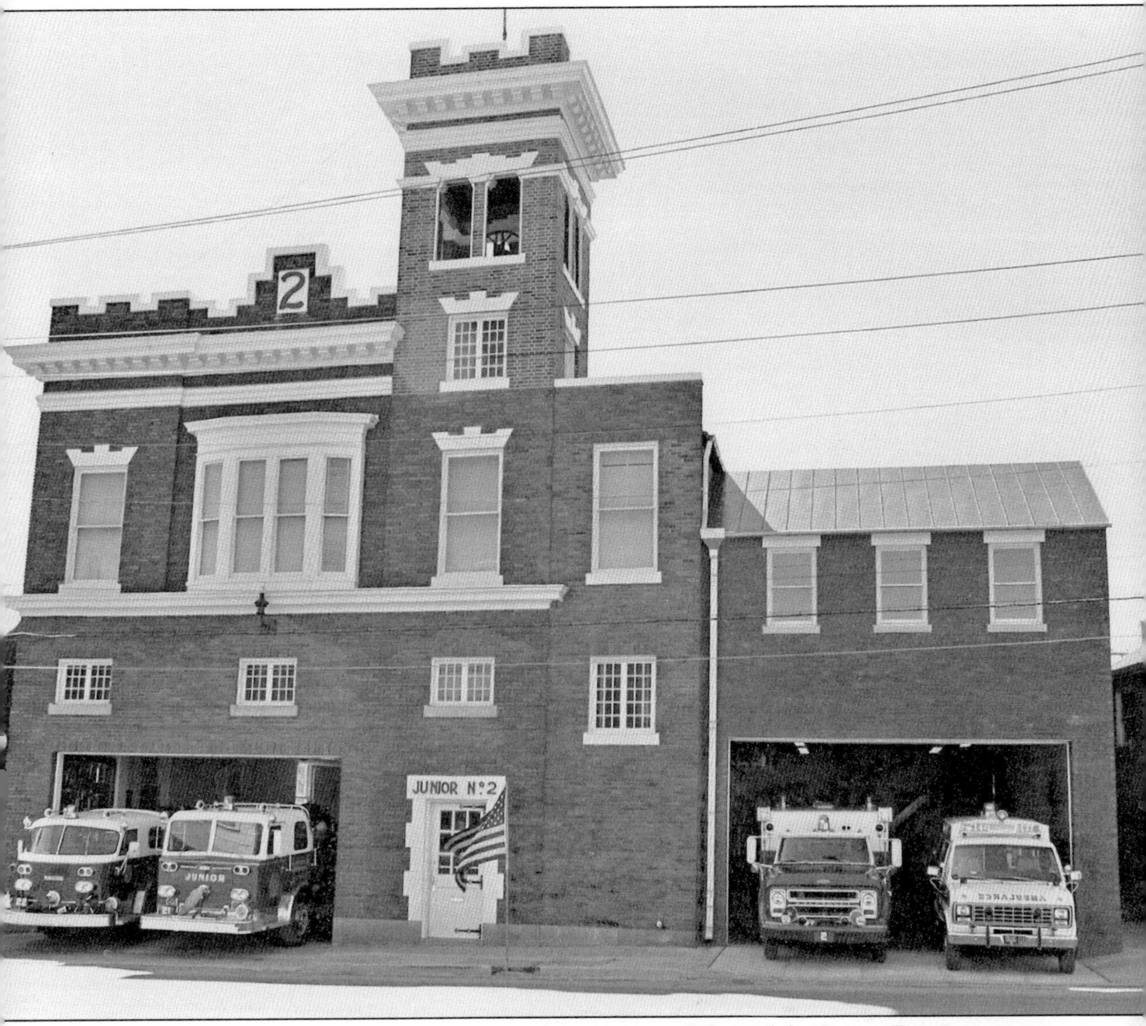

Apparatus is displayed during the dedication of the new north bay of the Junior Fire Station at 535 North Market Street in 1982. The new addition cost $242,000.

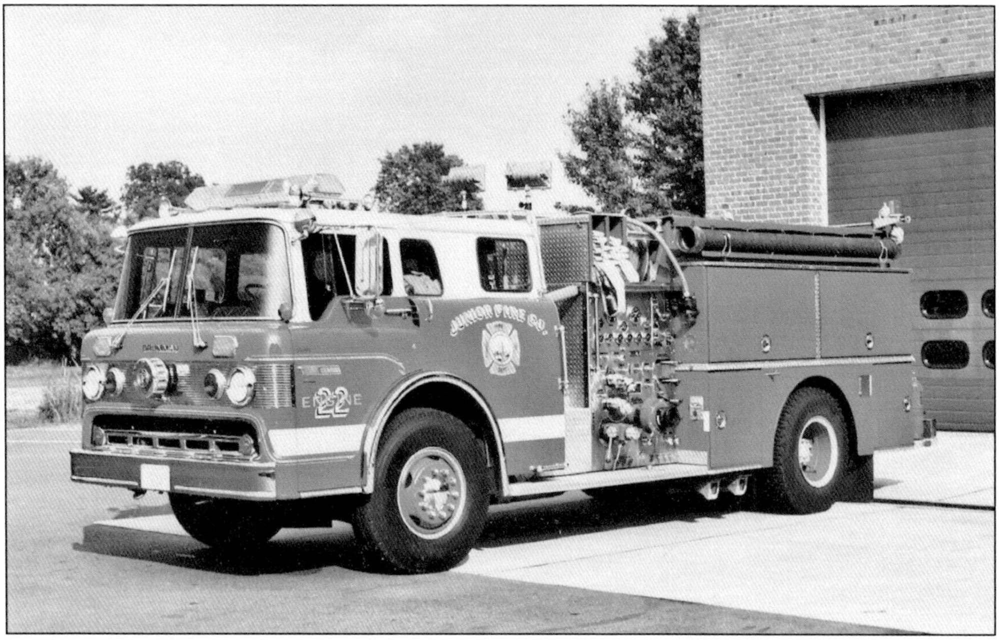

On September 9, 1986, the Juniors placed in service a 1986 Ford Grumman pumper. Company funds purchased this unit to provide an engine to respond to the growing county area surrounding Frederick.

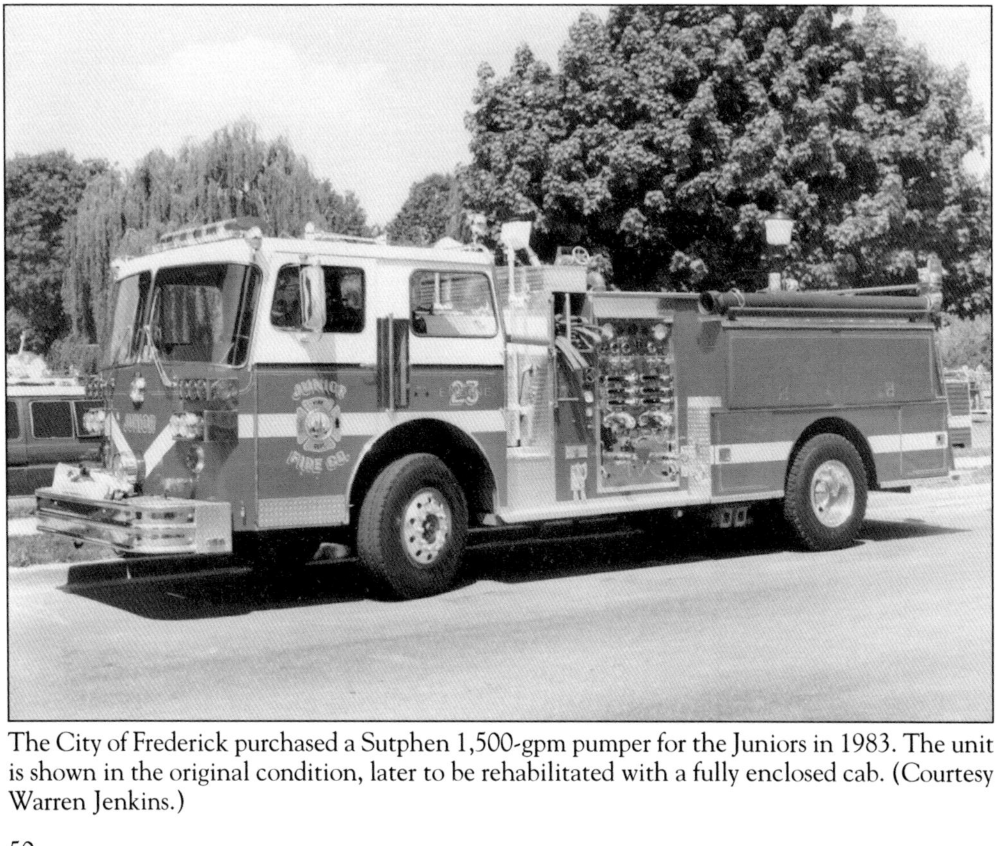

The City of Frederick purchased a Sutphen 1,500-gpm pumper for the Juniors in 1983. The unit is shown in the original condition, later to be rehabilitated with a fully enclosed cab. (Courtesy Warren Jenkins.)

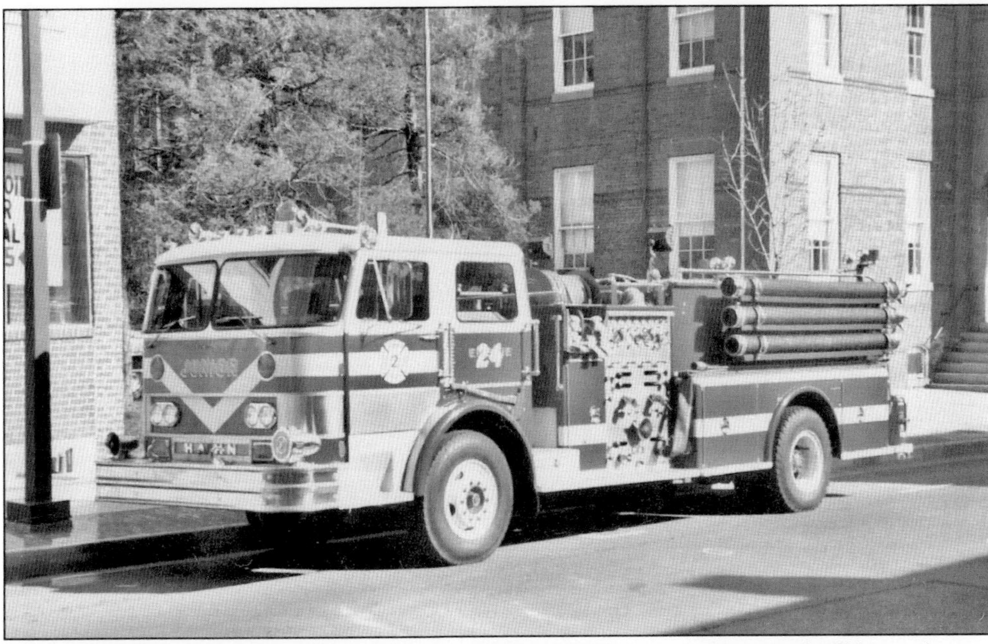

The Junior Fire Company raised money in 1985 to purchase a pumper with additional water supply for rural response. The 1973 Hahn was originally a seven-speed manual transmission diesel pumper from Hockessin, Delaware. (Courtesy Warren Jenkins.)

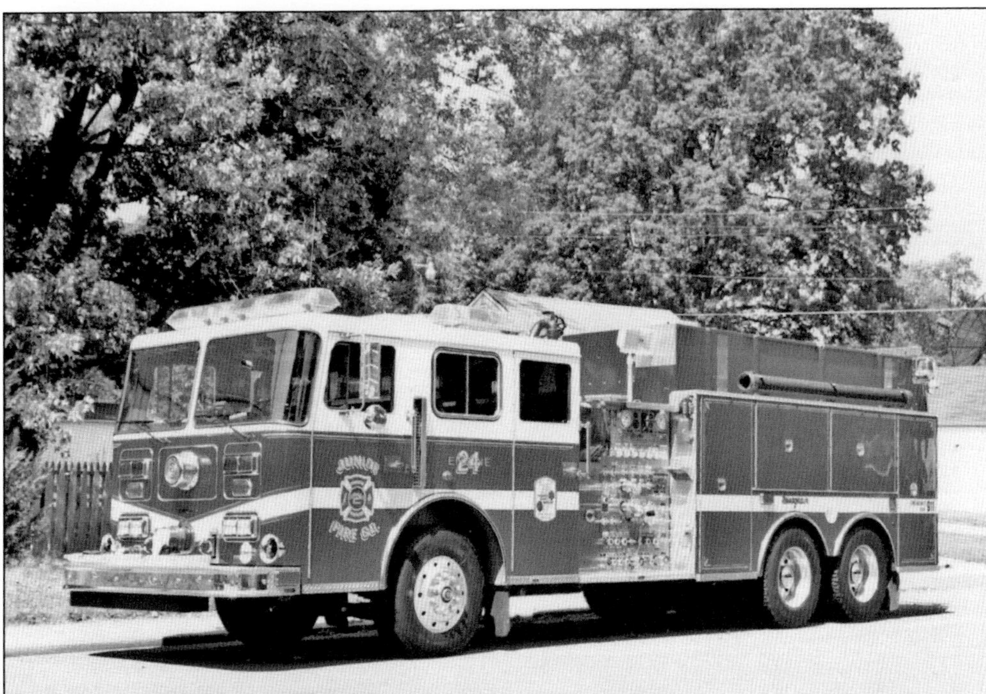

As the county area continued to grow, the Juniors received a Seagrave 2,000-gallon engine/tanker to help meet the demands of the rural areas.

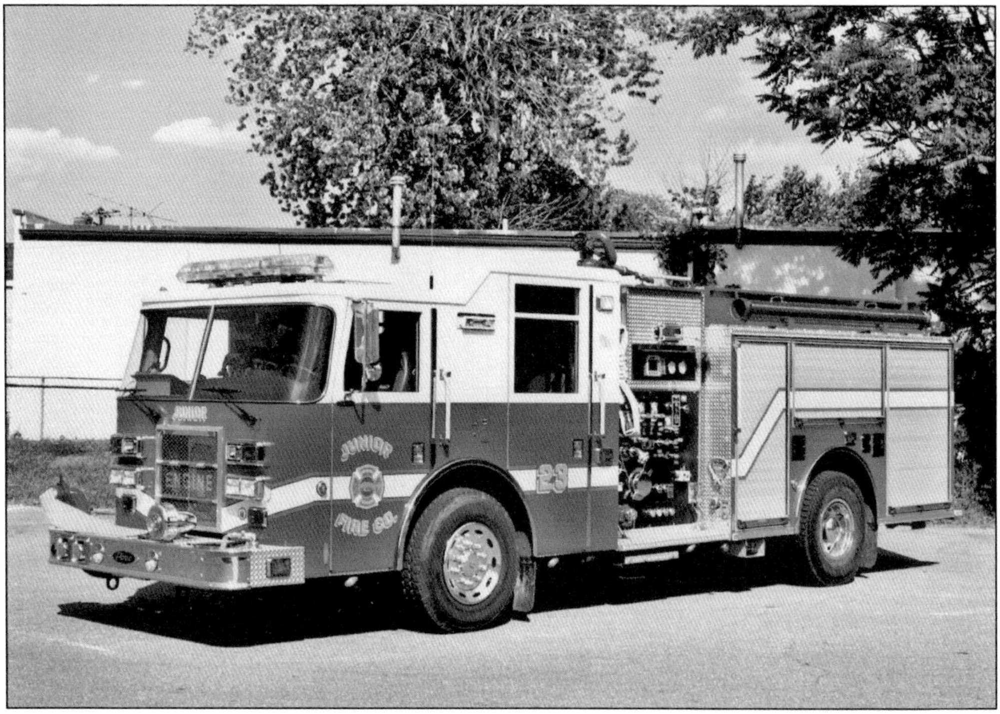

A recent edition to the Juniors fleet is the 2003 Pierce pumper. The unit replaced the 1983 Sutphen that now sees reserve service throughout the county. (Courtesy Warren Jenkins.)

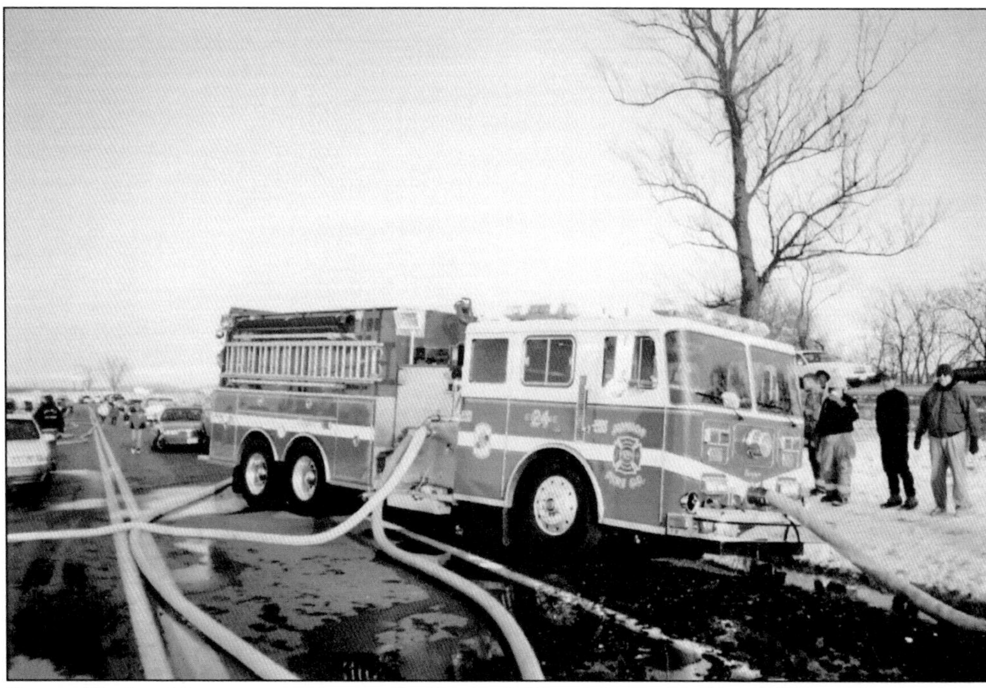

Engine/Tanker 24 pumps capacity at the Seventh Day Adventist Church Fire in Frederick.

Three

UNITED STEAM FIRE ENGINE COMPANY NO. 3

The Swampers

In 1837, the Washington Hose Company was organized on West All Saints Street to provide protection to the growing south end of Frederick. Unable to maintain a membership, the company disbanded in 1844 and was reorganized in 1845 as the Mechanics Hose Company, later renamed The Uniteds. When land was purchased for a new engine house in the swampy area of Carroll Creek, the town folk dubbed the building "Swamp Hall" and the membership became known as "Swampers."

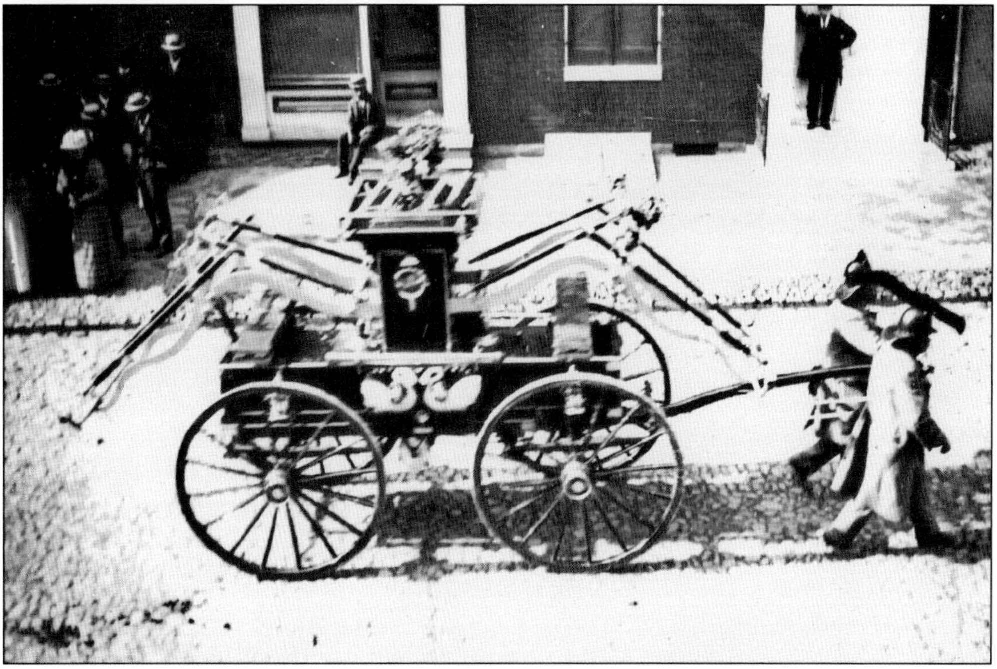

The first engine purchased by the United Fire Company was this 1815 Gallery hand pumper bought in 1851 from Baltimore. (Courtesy UFC.)

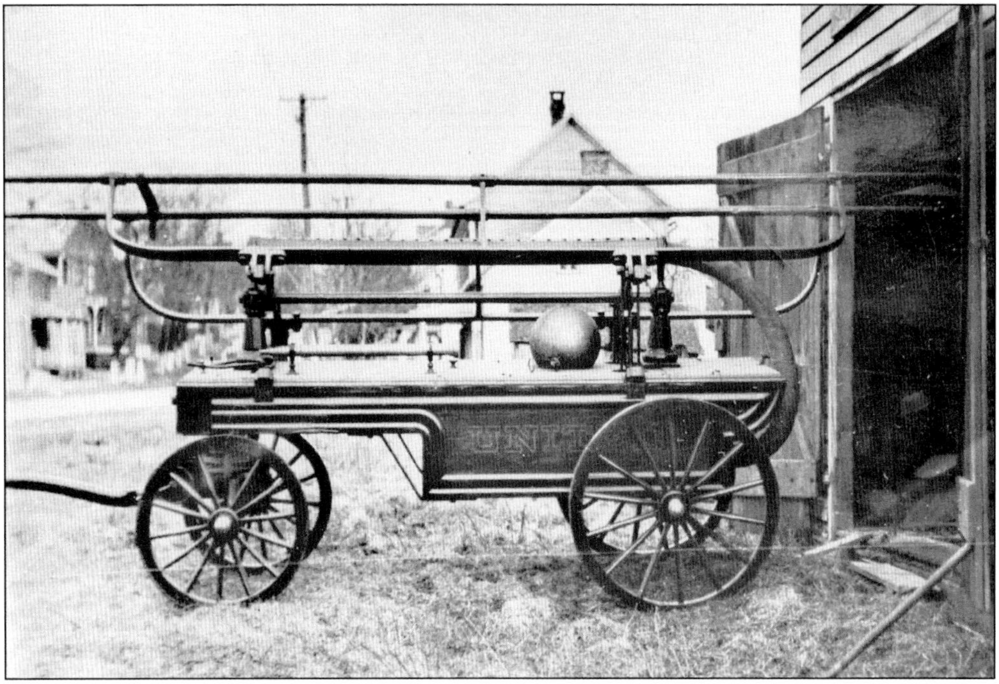

The "Old Lady" hand pumper was purchased used in 1860. It also had a companion trailer that dubbed this a squirrel-tail pumper. This pumper was sold to the Libertytown in the late 1800s. (Courtesy UFC.)

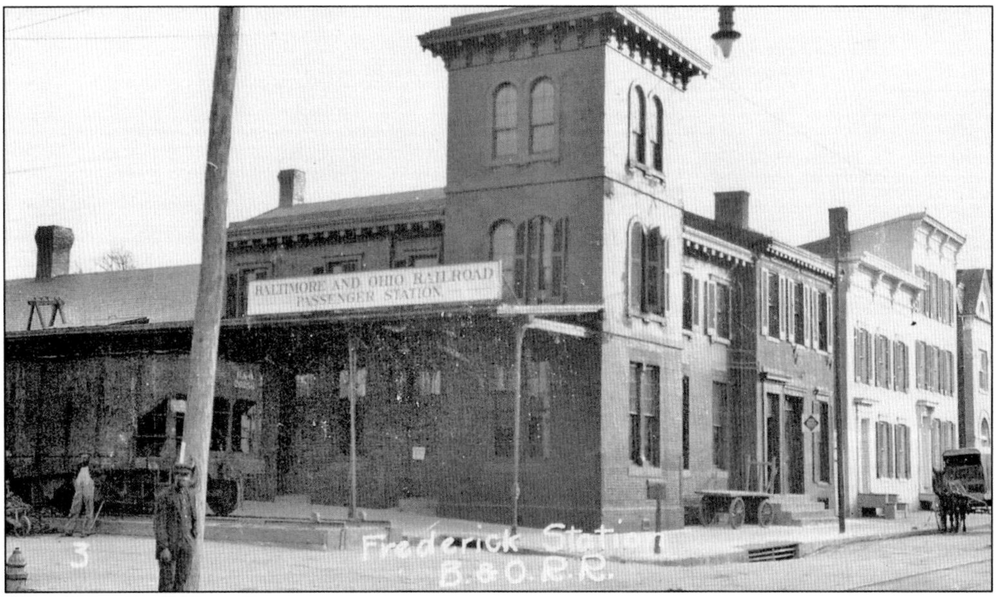

In 1858, Capt. John Sinn organized the United Guard, a local militia unit. The Independents organized the Independent Rifles, while the Juniors organized the Junior Defenders. These groups were summoned to Harpers Ferry in 1859 to assist in the John Brown insurrection. Each fire company militia marched to the B&O Railroad Station at Market and All Saints Street to board a special train. Upon arrival in Harpers Ferry, they assisted the federal troops on patrol functions. (Courtesy UFC.)

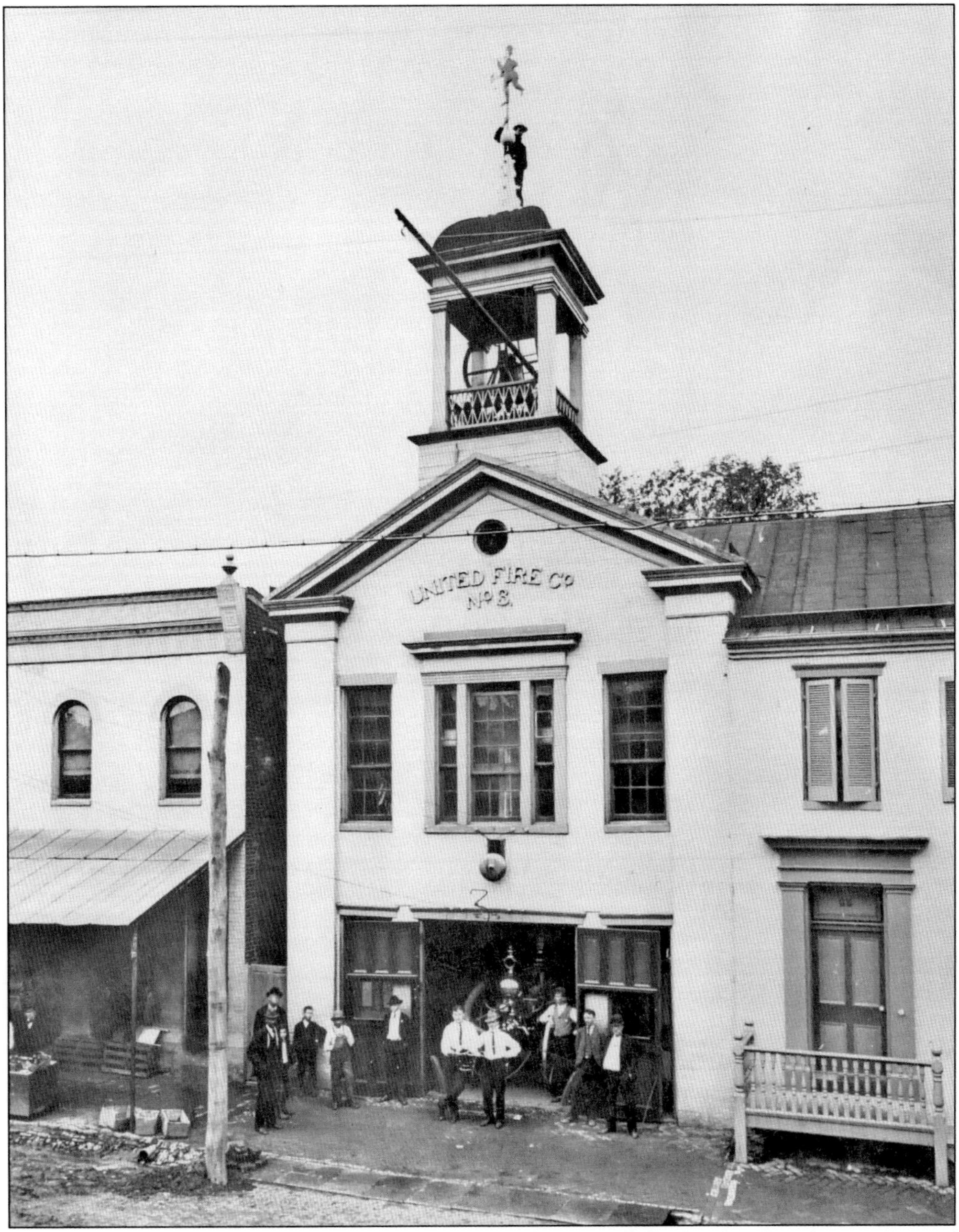

"Swamp Hall," the United Engine house, is shown here *c.* 1880, shortly after the purchase of the "Lily of the Swamp" steam pumper. (Courtesy UFC.)

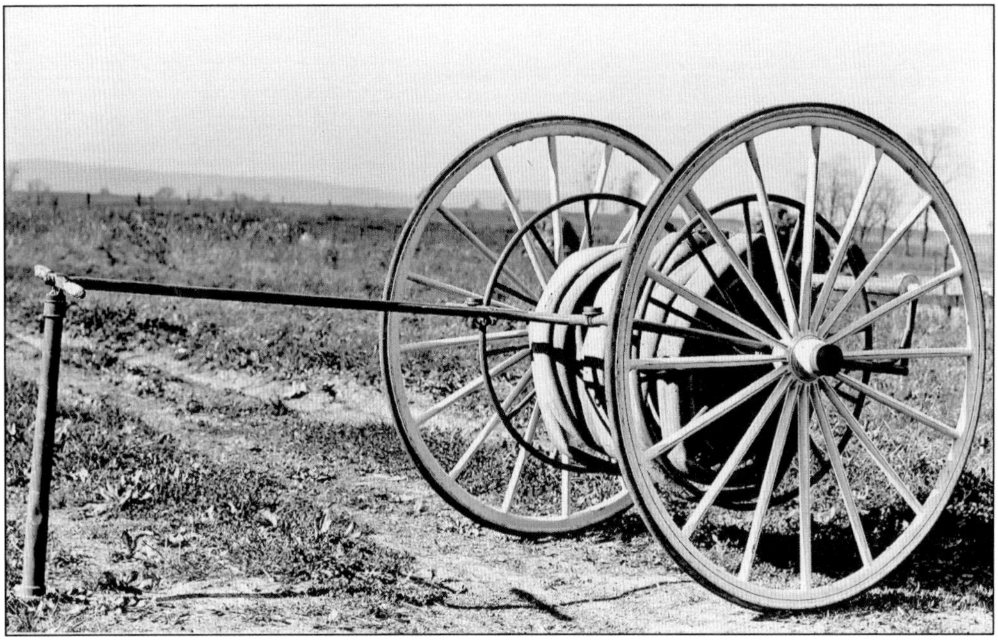

Around 1880, the Uniteds purchased hose reel to use with the chemical and steam pumper. (Courtesy UFC.)

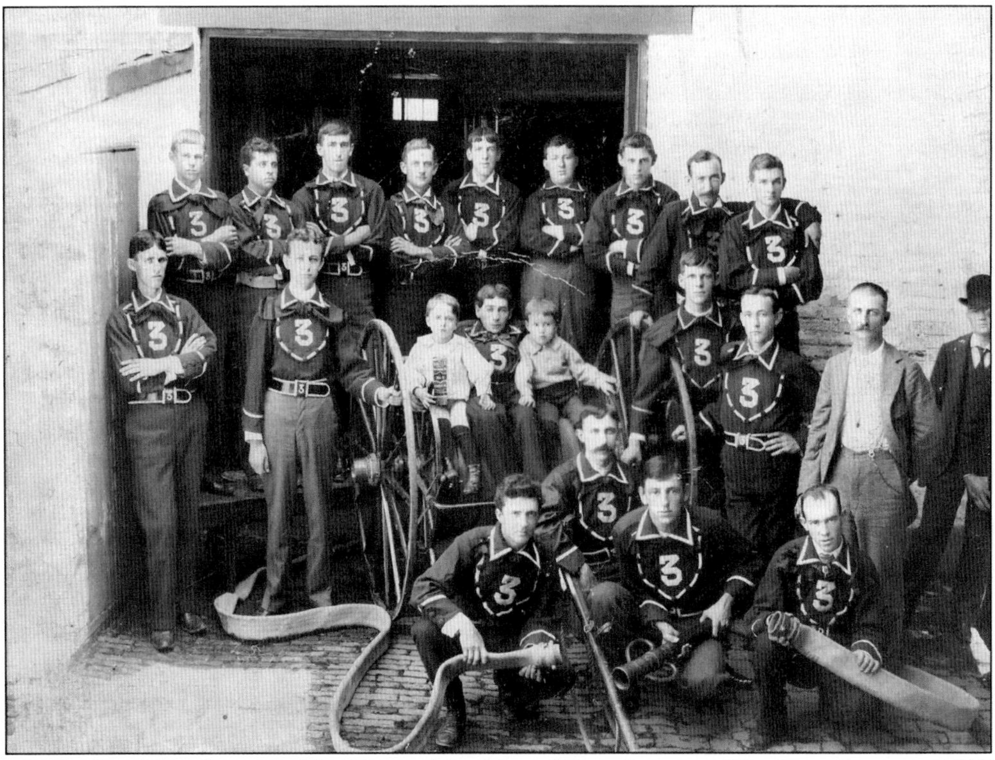

The Reel and Hook and Ladder Team of the United Fire Company No. 3 are shown in 1897 ready for competition. The team included many young Swampers and children of company members. Note several very young children on the reel (Courtesy UFC.)

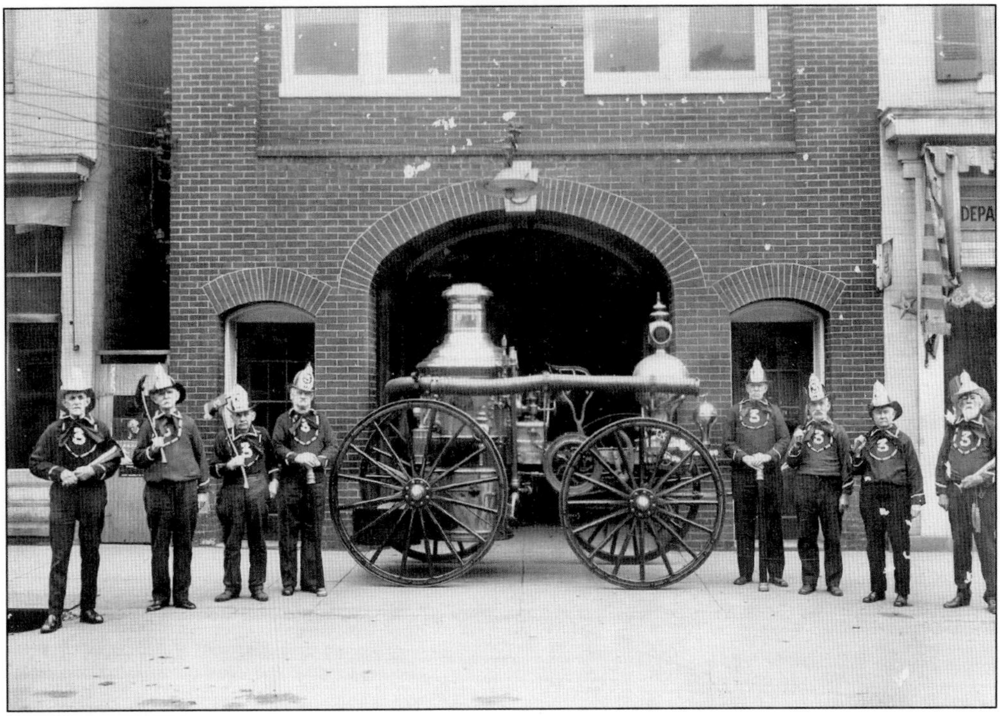

The famous "Lily of the Swamp," a Clapp & Jones engine, was donated to the Smithsonian Institute and was for many years on display in the Washington museum. The picture shows the Lily as they celebrate her 50th anniversary in 1928. The members shown actually helped place the engine in service in 1878. (Courtesy UFC.)

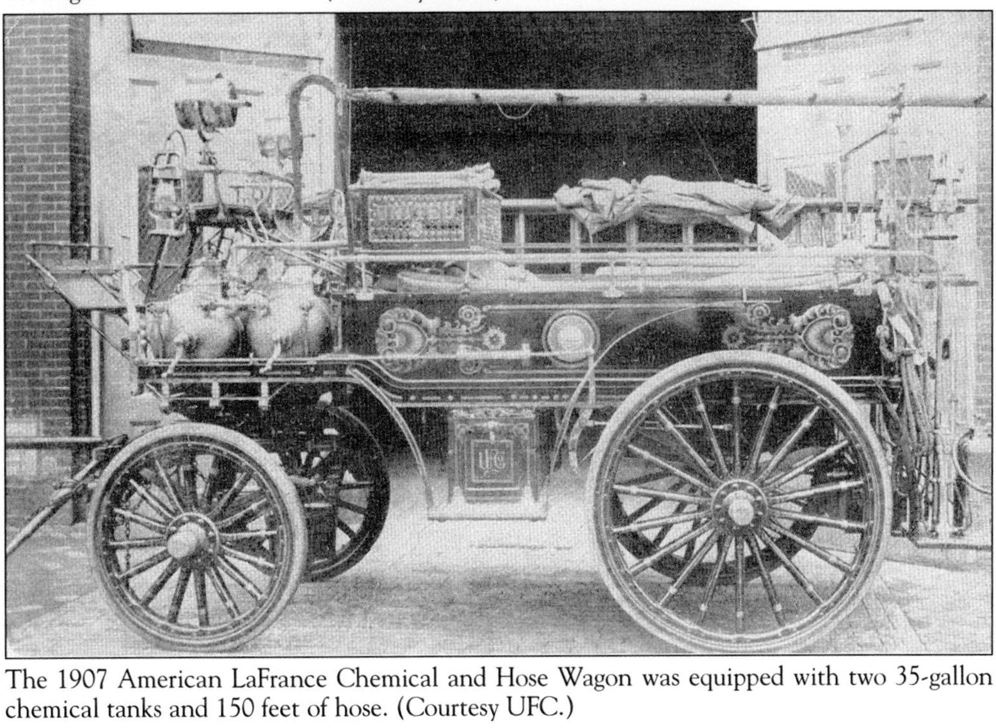

The 1907 American LaFrance Chemical and Hose Wagon was equipped with two 35-gallon chemical tanks and 150 feet of hose. (Courtesy UFC.)

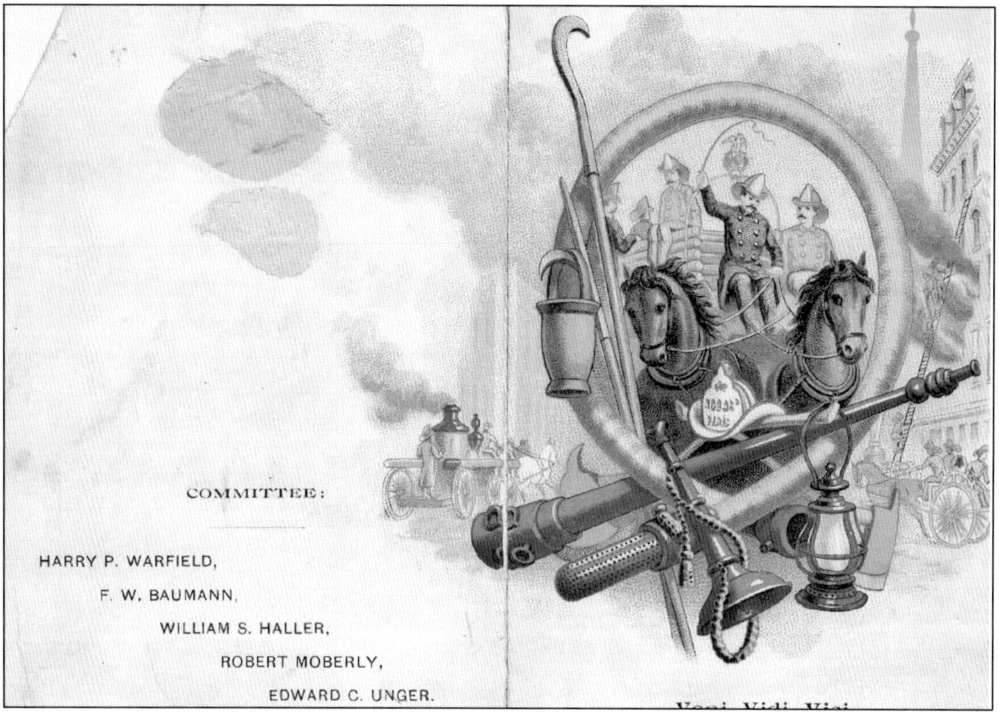

A copy of the 1890 Annual Ball invitation displays the company motto, "Veni, Vidi, Vici" (I Came, I Saw, I Conquered.) (Courtesy UFC.)

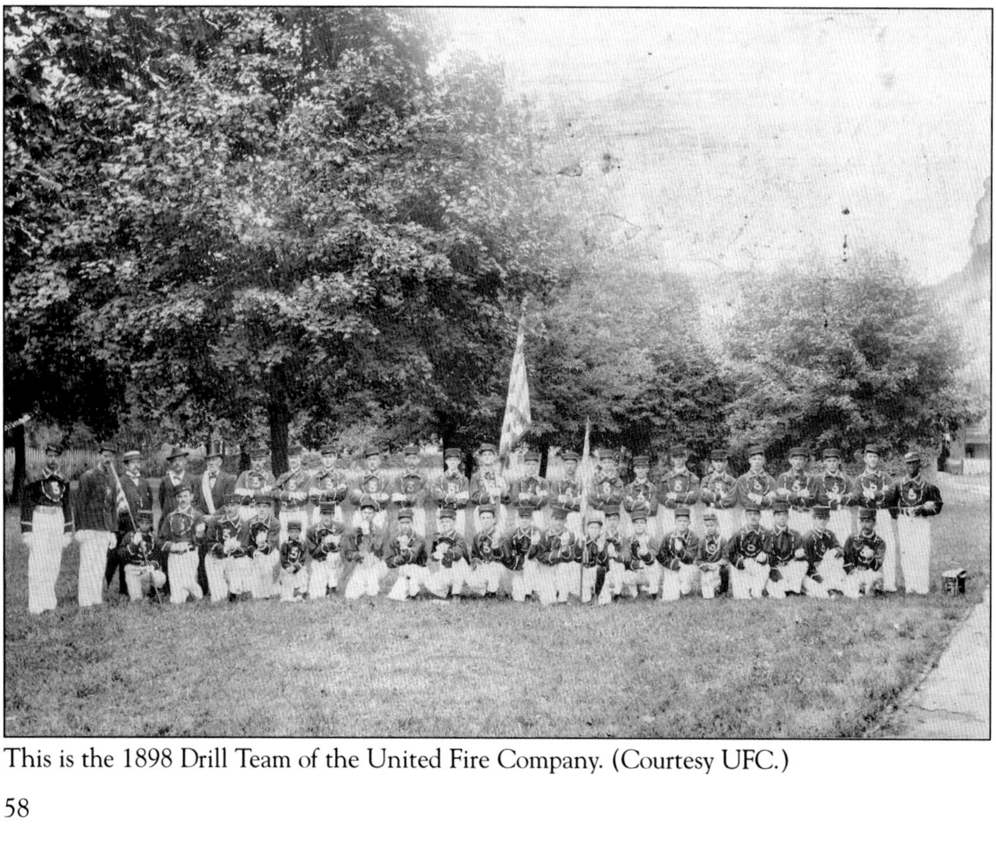

This is the 1898 Drill Team of the United Fire Company. (Courtesy UFC.)

Atop the "Swamp Hall," Johnny Swamper, the company mascot, stands vigilant watch over Frederick. (Courtesy UFC.)

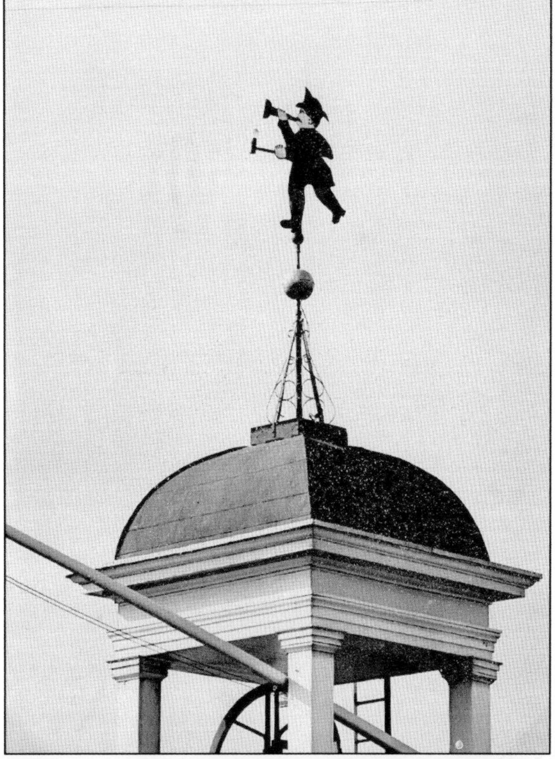

Close-up view of "Johnny Swamper," the weathervane trade mark of the United Steam Fire Engine Company No. 3.

Each fire company in Frederick participated in the bazaar at the City Opera House to raise money. This event, held in 1905, was used by the Uniteds to raise money to buy horses. (Courtesy UFC.)

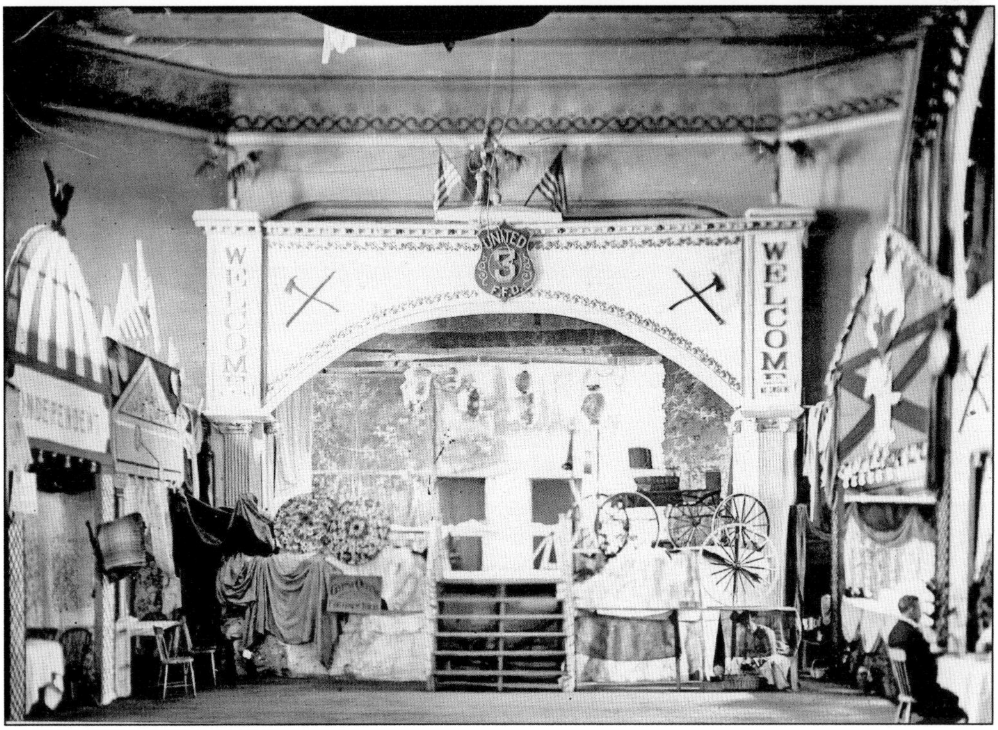

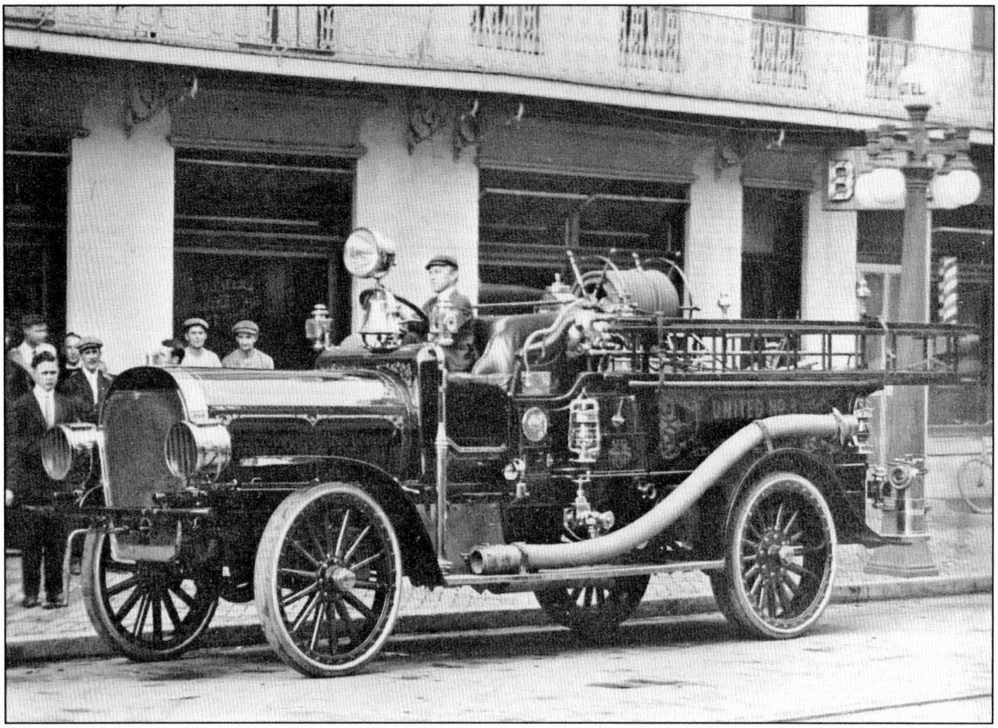

The United Waterous pumper sits in front of the City Hotel in 1912. (Courtesy UFC.)

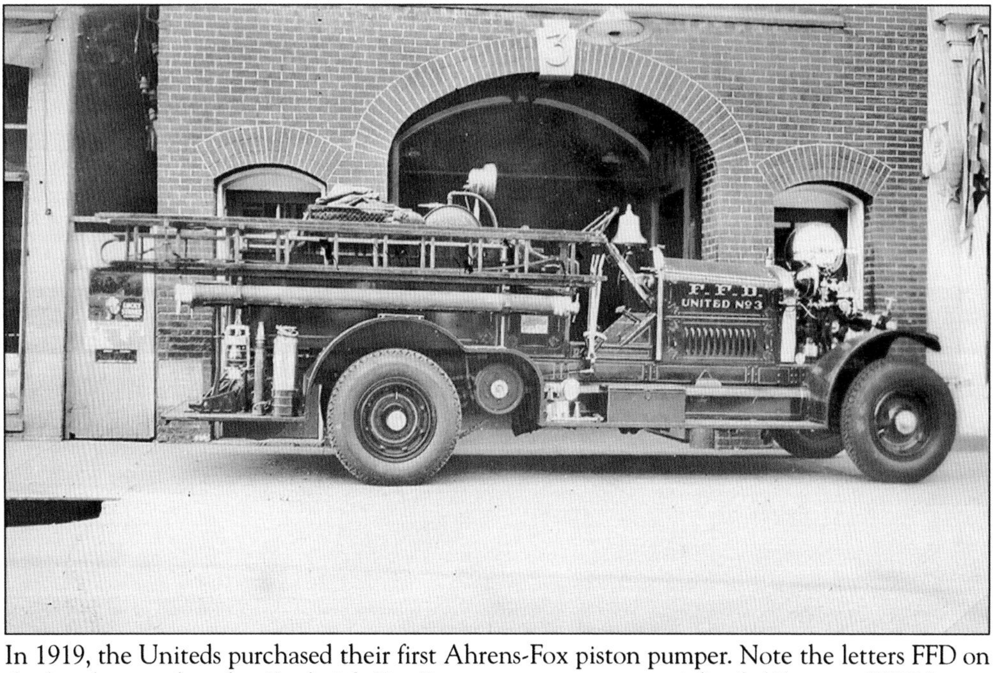

In 1919, the Uniteds purchased their first Ahrens-Fox piston pumper. Note the letters FFD on the hood, even though a Frederick Fire Department never materialized. (Courtesy UFC.)

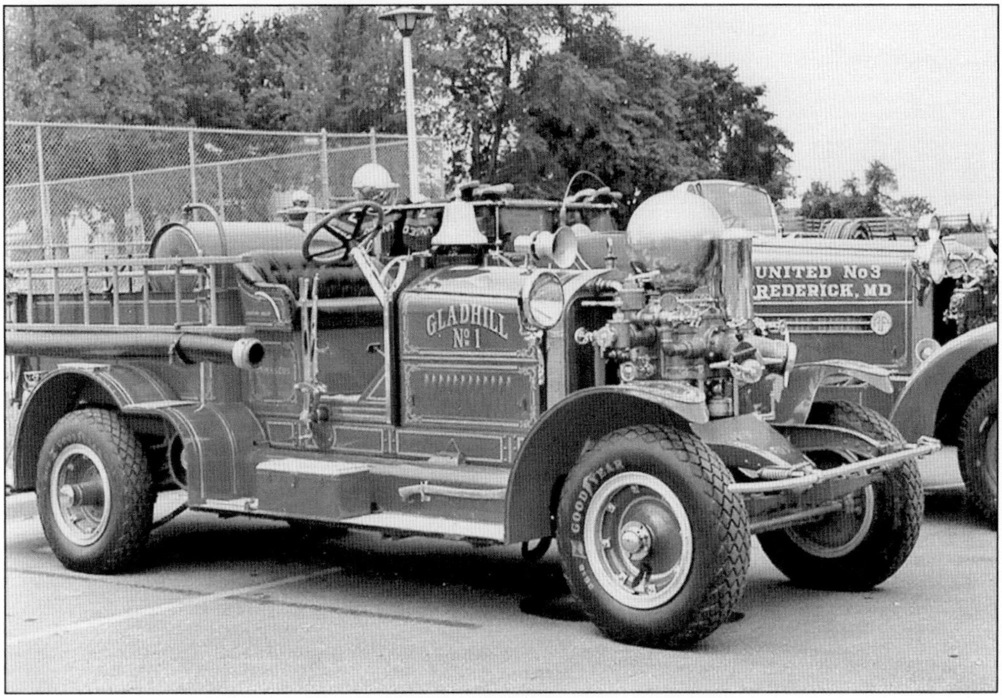

The 1919 Fox was purchased by Mr. Buck Gladhill of Damascus, who restored and repainted the engine Gladhill No. 1. (Courtesy H. Richard Hahn.)

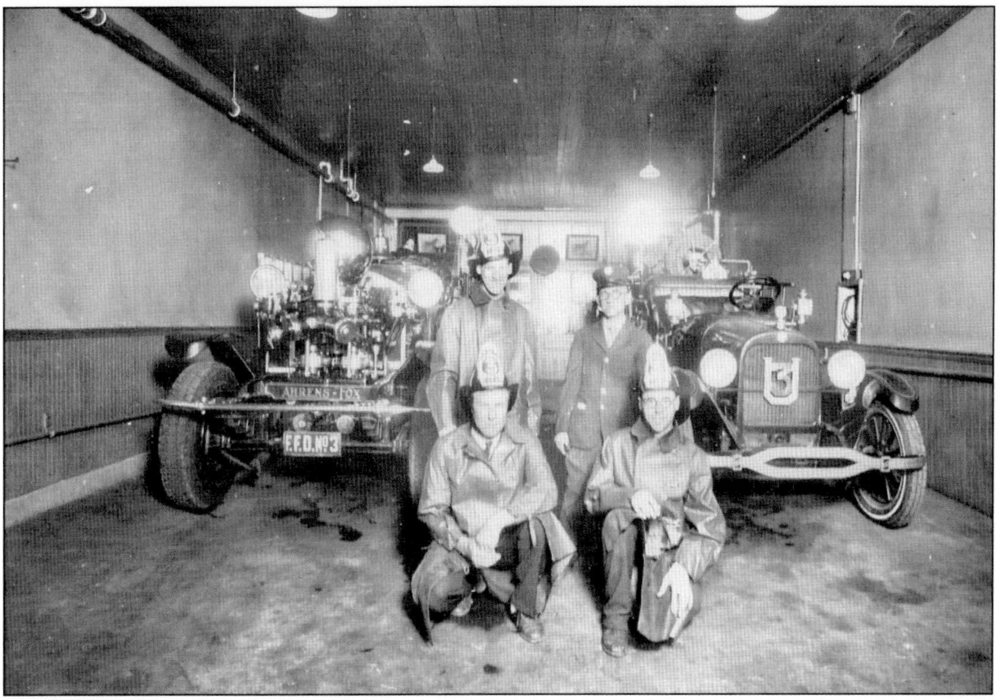

United members M. Unglebower, Ralph Baker, Carl Brown, and Charles Lipps pose in front of the 1919 Ahrens-Fox and 1923 floodlight truck. (Courtesy UFC.)

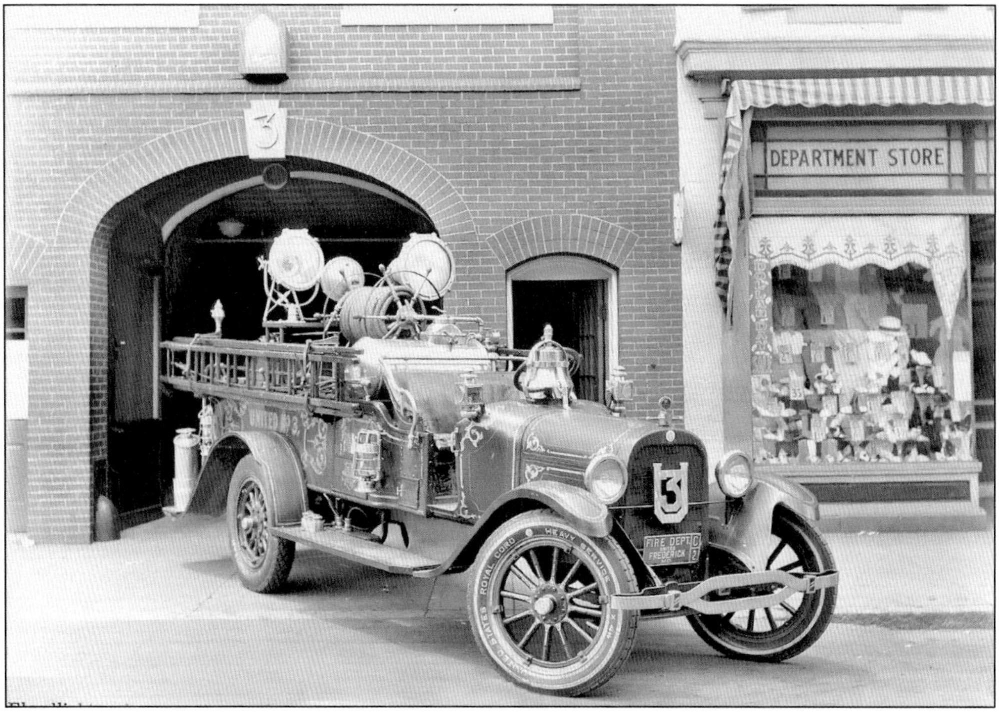

The first floodlight and rescue truck was actually the 1912 Waterous body attached to a 1923 Dodge chassis. (Courtesy UFC.)

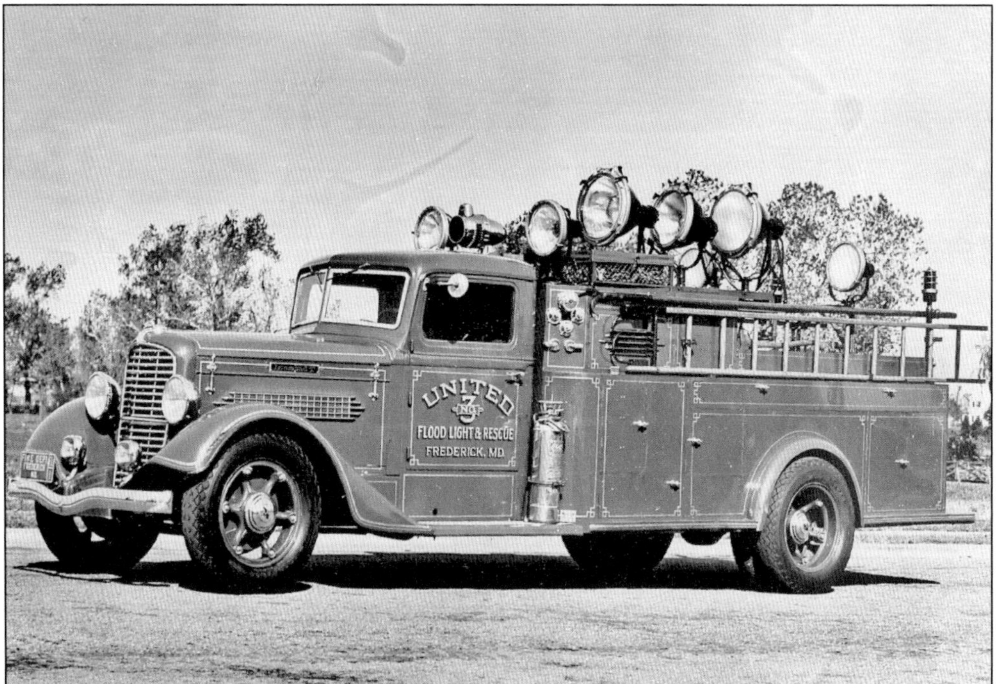

The 1923 Dodge was replaced by a 1935 Diamond T floodlight and rescue truck designed by member "Billy" Lipps. (Courtesy UFC.)

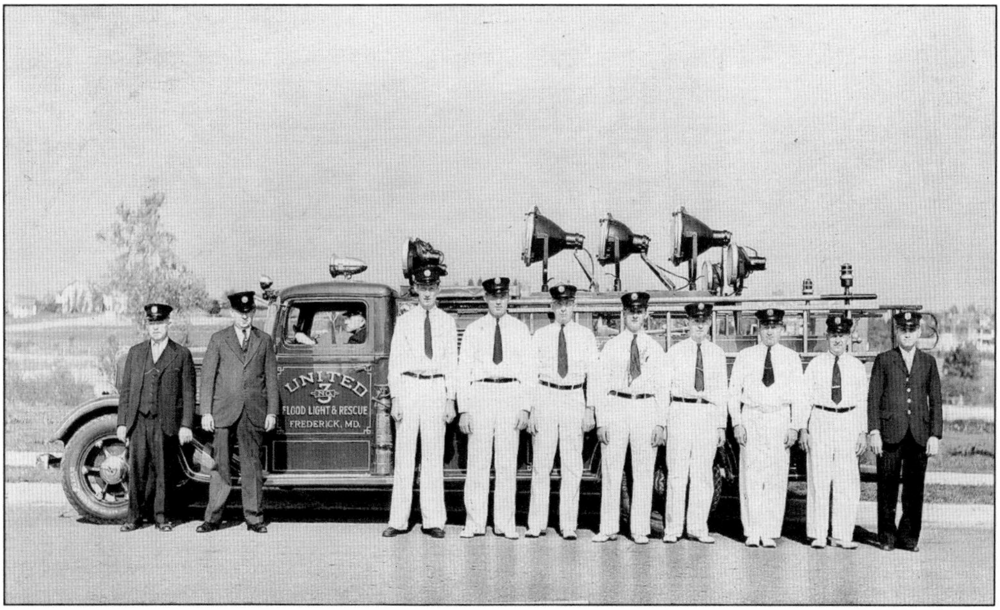

The Uniteds organized the first rescue squad to respond on the new rescue truck. Members, from left to right, are William D. Lipps, W. Bartgis Storm, Charles E. Lipps, John Engelbrecht, Arthur "Kelly" Hoffman, Arthur C. Aldridge, Charles Marman, Alton Lipps, Charles E. Kinna, Walter Duvall, and Dr. James Marrone (captain). (Courtesy UFC.)

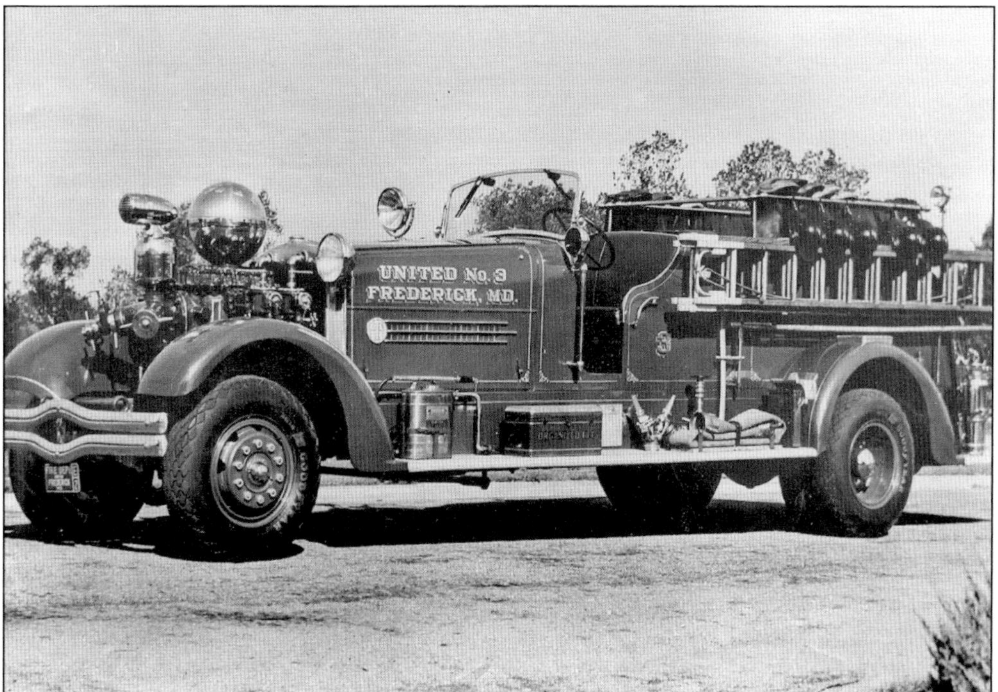

The Uniteds continued the tradition of Ahrens-Fox by purchasing the 1939 750-gpm piston pumper. This unit remained in service until July 1, 1980, and is still popular at parades and musters. (Courtesy UFC.)

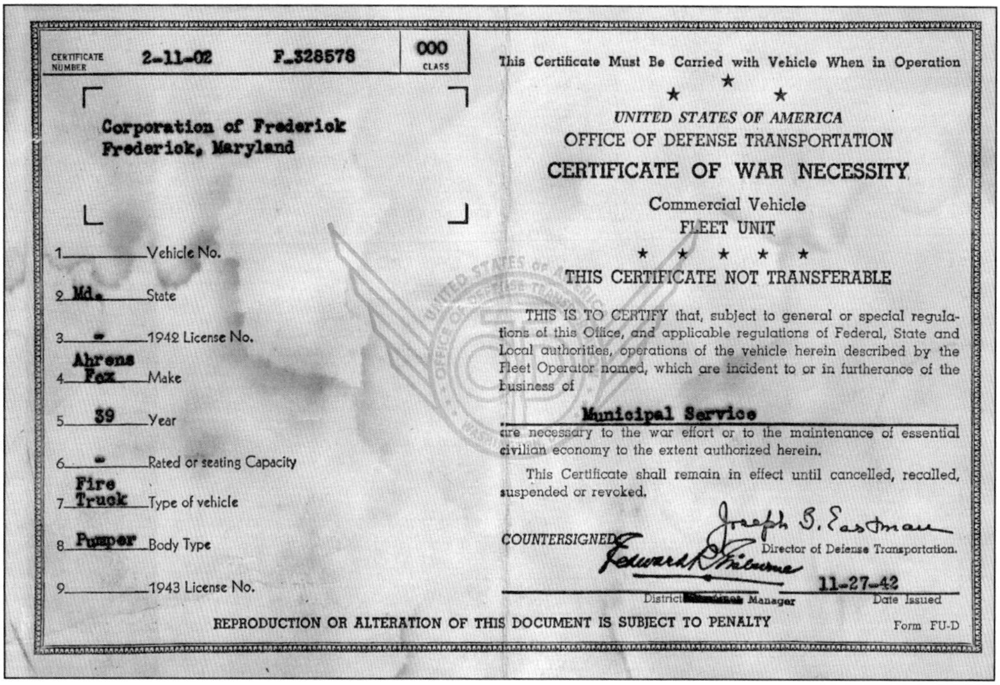

CERTIFICATE NUMBER 2-11-02 F-328578 000 CLASS

Corporation of Frederick
Frederick, Maryland

1.______Vehicle No.

2. Md. State

3. * 1942 License No.

4. Ahrens Fox Make

5. 39 Year

6. * Rated or seating Capacity

7. Fire Truck Type of vehicle

8. Pumper Body Type

9.______1943 License No.

This Certificate Must Be Carried with Vehicle When in Operation

★ ★ ★

UNITED STATES OF AMERICA

OFFICE OF DEFENSE TRANSPORTATION

CERTIFICATE OF WAR NECESSITY

Commercial Vehicle

FLEET UNIT

★ ★ ★ ★ ★

THIS CERTIFICATE NOT TRANSFERABLE

THIS IS TO CERTIFY that, subject to general or special regulations of this Office, and applicable regulations of Federal, State and Local authorities, operations of the vehicle herein described by the Fleet Operator named, which are incident to or in furtherance of the business of

Municipal Service

are necessary to the war effort or to the maintenance of essential civilian economy to the extent authorized herein.

This Certificate shall remain in effect until cancelled, recalled, suspended or revoked.

Joseph B. Eastman
Director of Defense Transportation.

COUNTERSIGNED Edward R. Milcume
District Manager

11-27-42
Date Issued

REPRODUCTION OR ALTERATION OF THIS DOCUMENT IS SUBJECT TO PENALTY

Form FU-D

All municipal vehicles had to receive a special certificate during World War II to show it was necessary to "maintenance of essential civilian economy." (Courtesy UFC.)

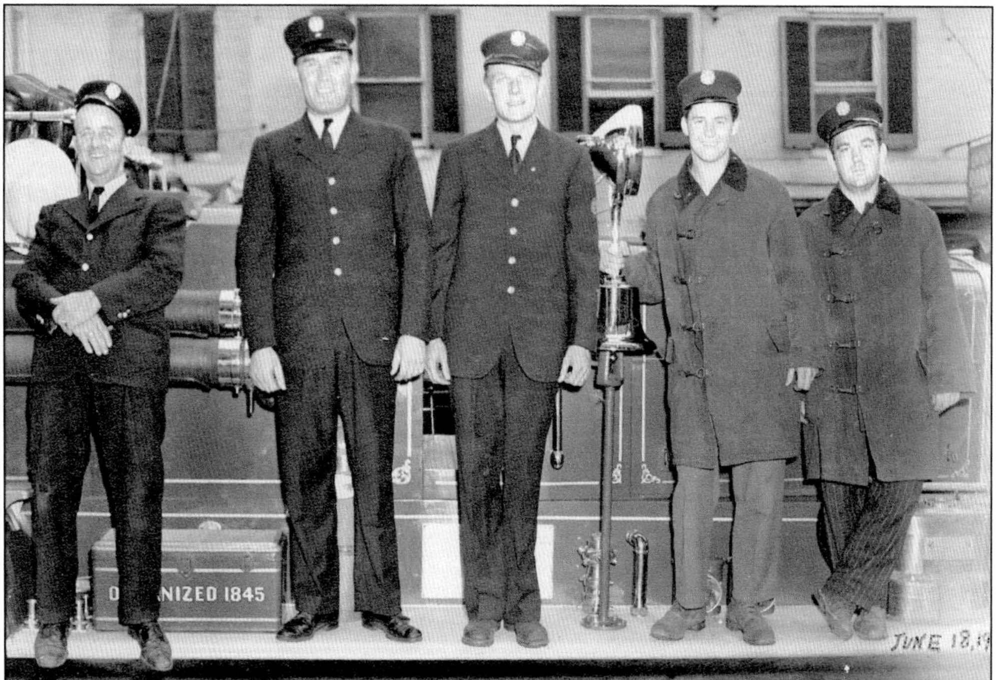

Standing by for a run at the United Firehouse, from left to right, are John Bowers, Arthur "Kelly" Hoffman, Joe Hooper, Charles E. "Moose" Miller, and Thomas Carbaugh on June 18, 1946. (Courtesy UFC.)

In 1954, the United Steam Fire Engine Company No. 3 purchased a new Cadillac ambulance and began the first fire department ambulance service in Frederick. (Courtesy UFC.)

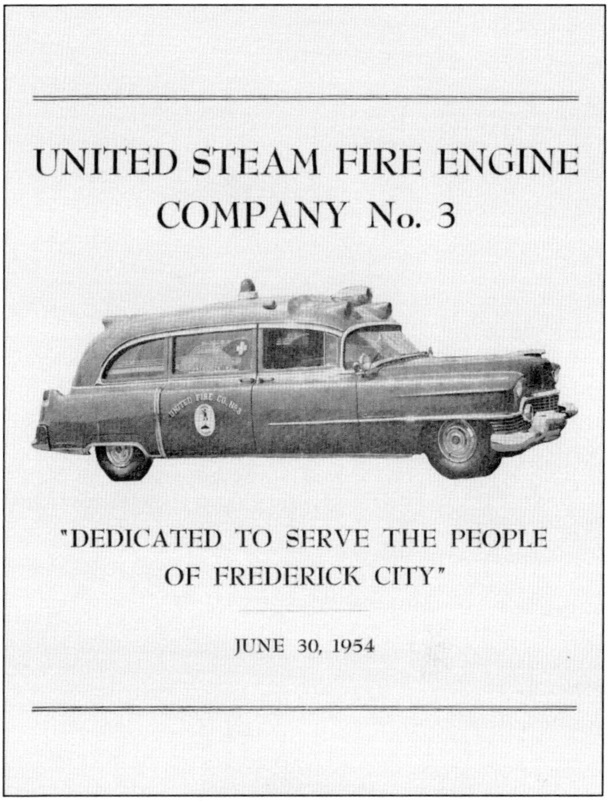

UNITED STEAM FIRE ENGINE COMPANY No. 3

"DEDICATED TO SERVE THE PEOPLE OF FREDERICK CITY"

JUNE 30, 1954

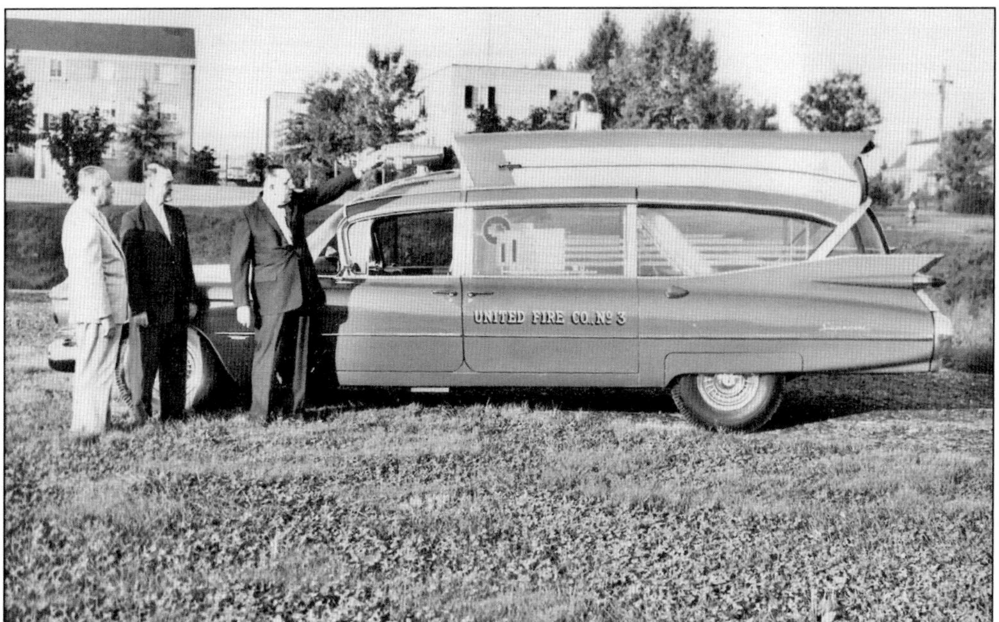

In 1959, a second Cadillac ambulance was purchased. Shown inspecting the new vehicle, from left to right, are Dr. James Marrone, Mayor Jacob R. Ramsburg, and United president Alton Y. Bennett. (Courtesy UFC.)

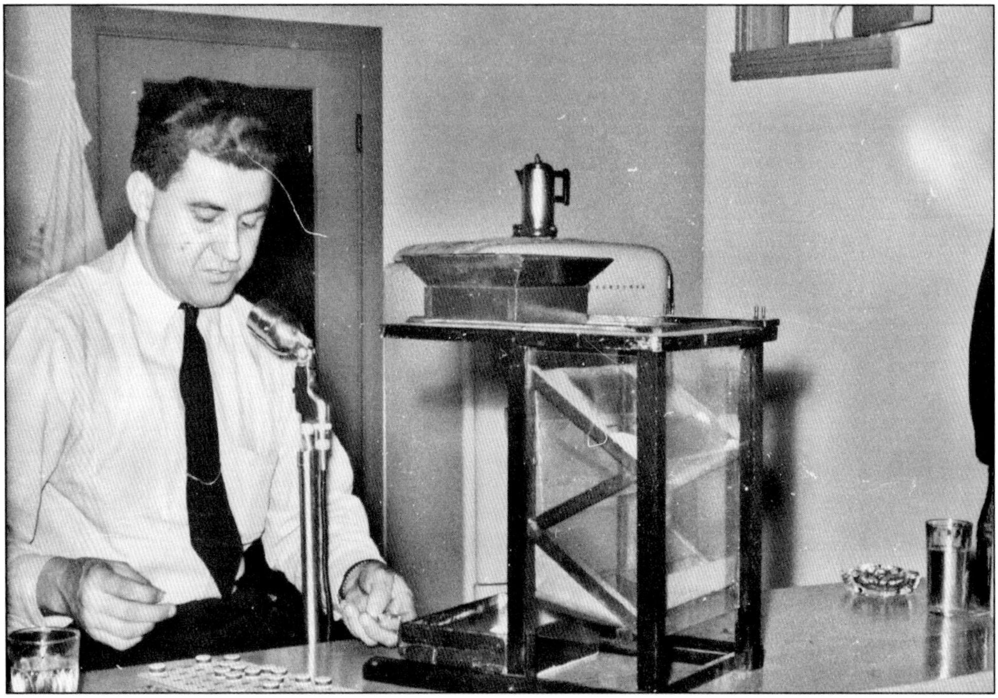

To raise money for the ambulance, the Uniteds sponsored "penny bingo" at the United Fire Hall. Ambulance driver Charles "Moose" Miller calls the numbers. (Courtesy UFC.)

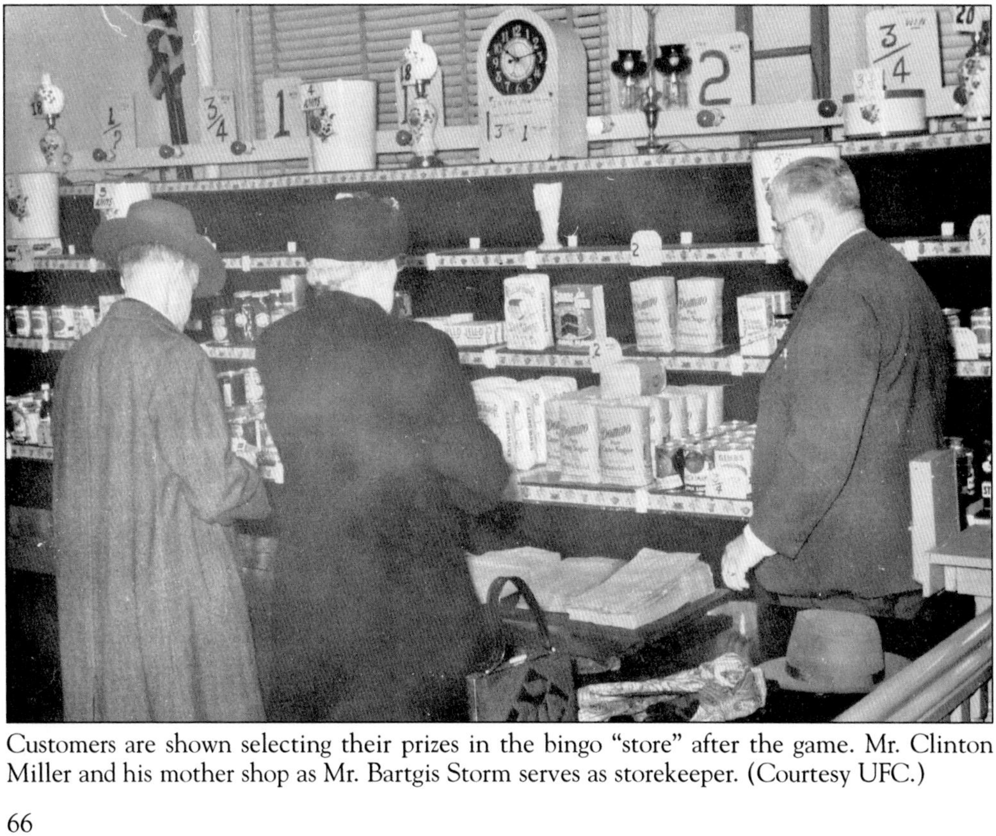

Customers are shown selecting their prizes in the bingo "store" after the game. Mr. Clinton Miller and his mother shop as Mr. Bartgis Storm serves as storekeeper. (Courtesy UFC.)

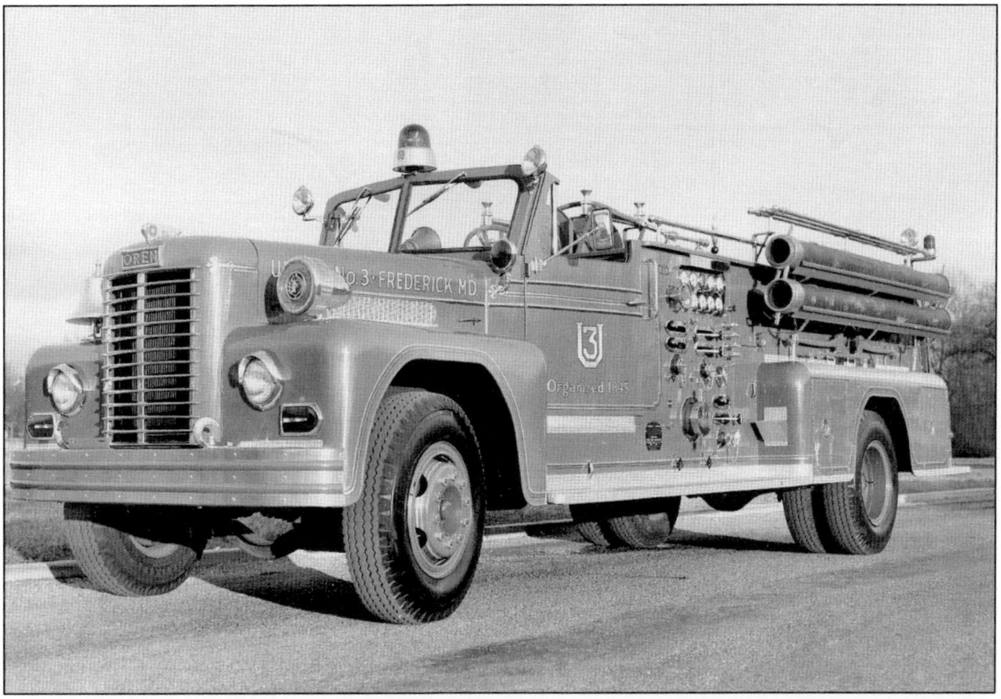

The last open cab engine in Frederick was the 1958 Oren purchased by the Uniteds. It was retired from service in 1970. (Courtesy UFC.)

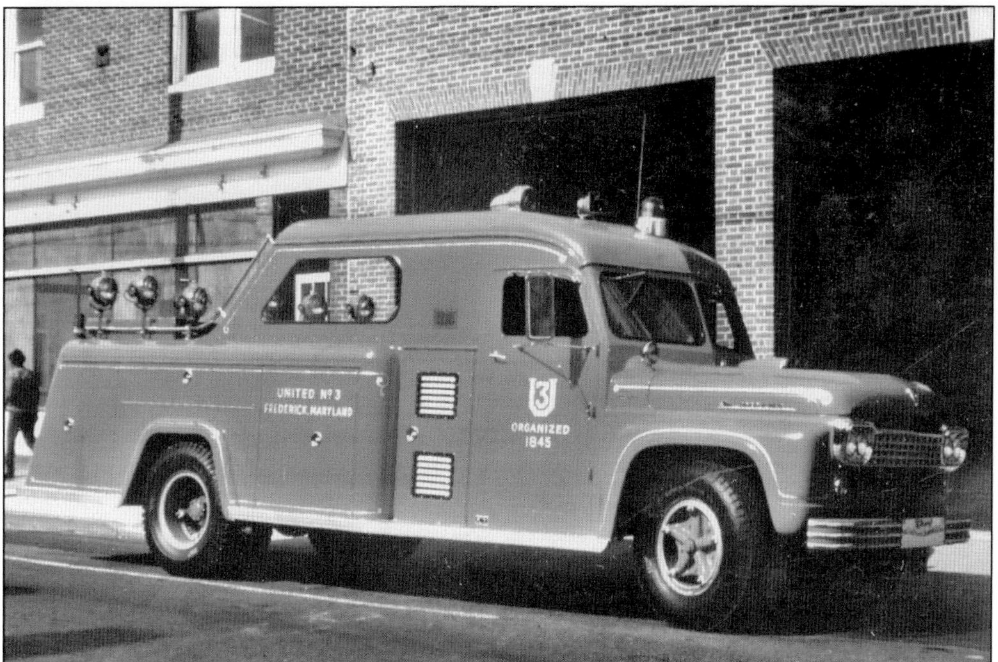

The Diamond T rescue truck was replaced in 1959 with a Ford chassis equipped with a Young rescue body. This unit was the first vehicle equipped with an electronic siren. (Courtesy Warren Jenkins.)

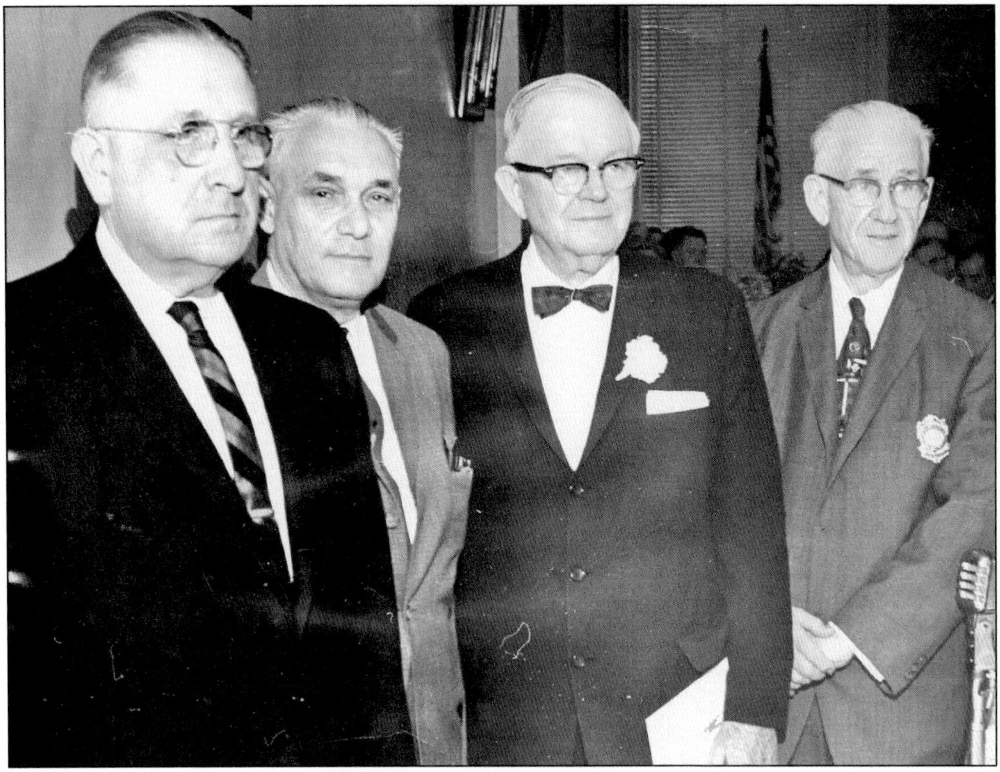

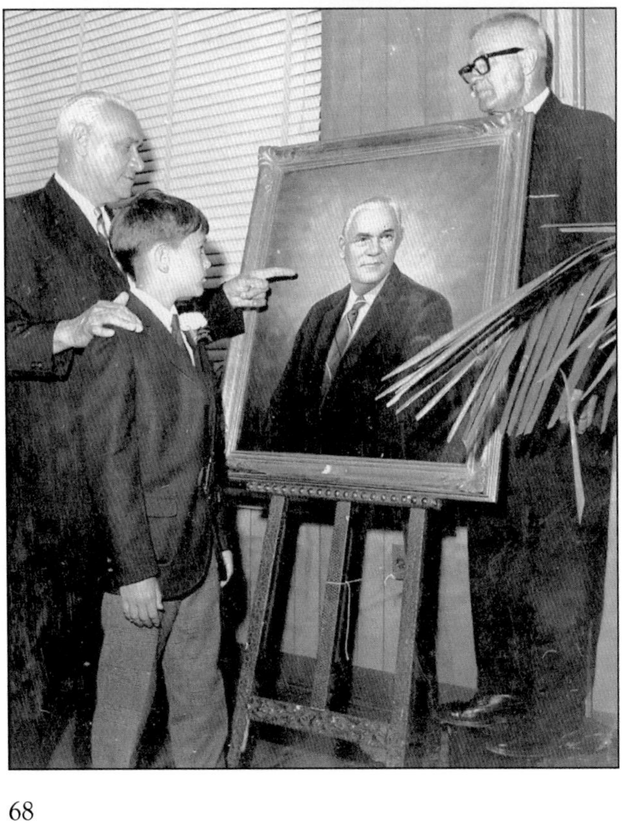

Alton Y. Bennett was honored at a dinner in May 1962 for his many years of service to the Uniteds and the community. This picture shows, from left to right, Mr. Bennett, Dr. James Marrone, Gov. Millard Tawes, and W. Bartgis Storm. (Courtesy UFC.)

Dr. James Marrone was presented a portrait at a dinner in his honor on May 6, 1969, for his service as president and father of the ambulance service. Shown are Dr. Marrone and his grandson James (left) and Ellis Carty (right). (Courtesy UFC.)

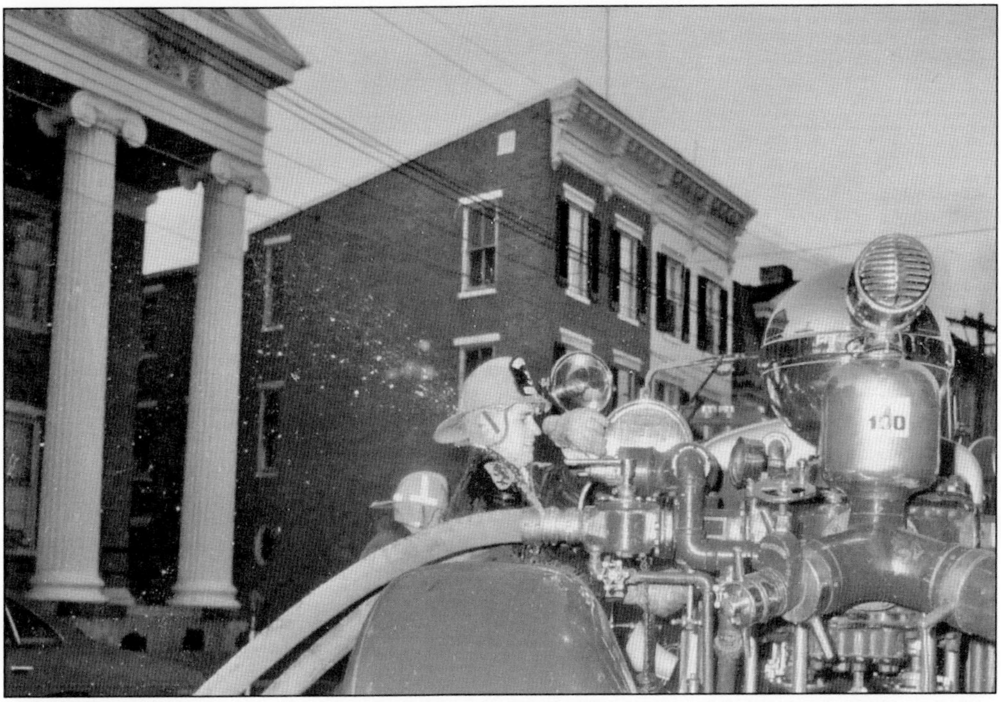

The 1939 Ahrens-Fox is shown in action at a working fire at the old YMCA on West Church Street. The pumper is hooked to a hydrant near Winchester Hall on East Church Street. The driver is Sgt. Herman Claggett. (Courtesy UFC.)

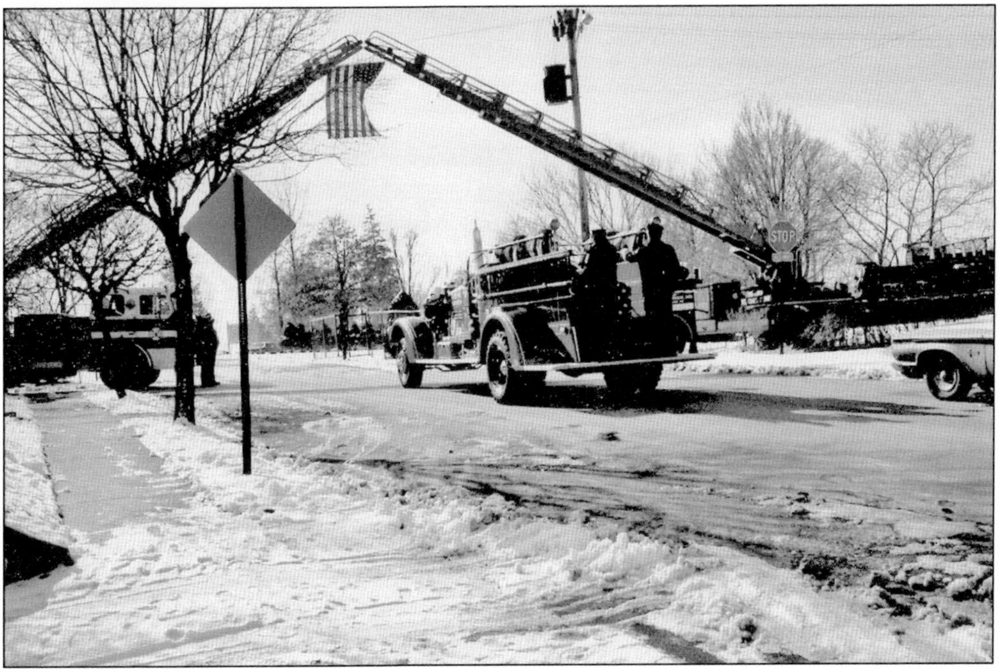

One of the last official duties of the 1939 Ahrens-Fox was to take Sgt. Charles E. "Moose" Miller on his final alarm. Moose was the caretaker of this engine for many years. (Collection of Charles M. Hahn.)

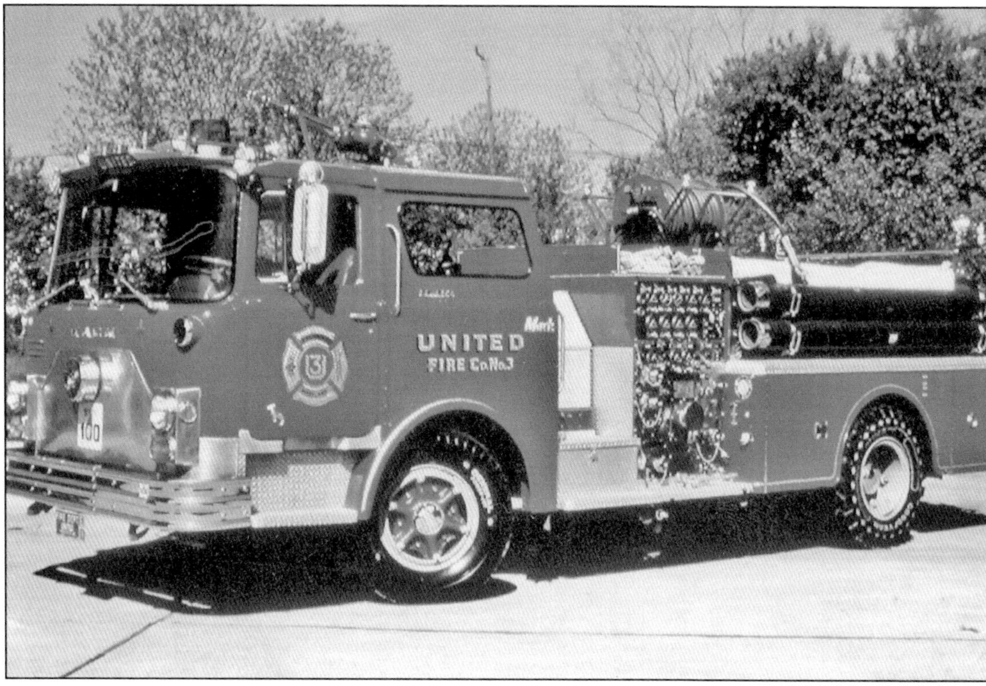

The Uniteds placed the first diesel fire engine in service in 1971 with the purchase of the CF Series Mack pumper. (Courtesy Warren Jenkins.)

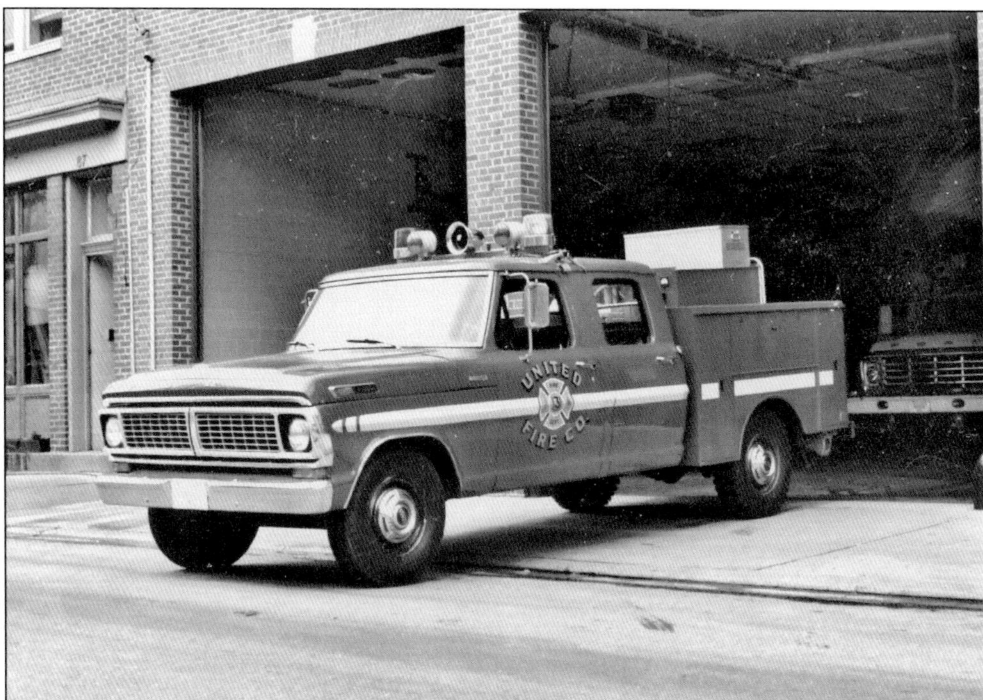

Utility 3 served as a foam unit carrying a high-expansion unit. It was used primarily for aircraft and petroleum fires. (Courtesy Warren Jenkins.)

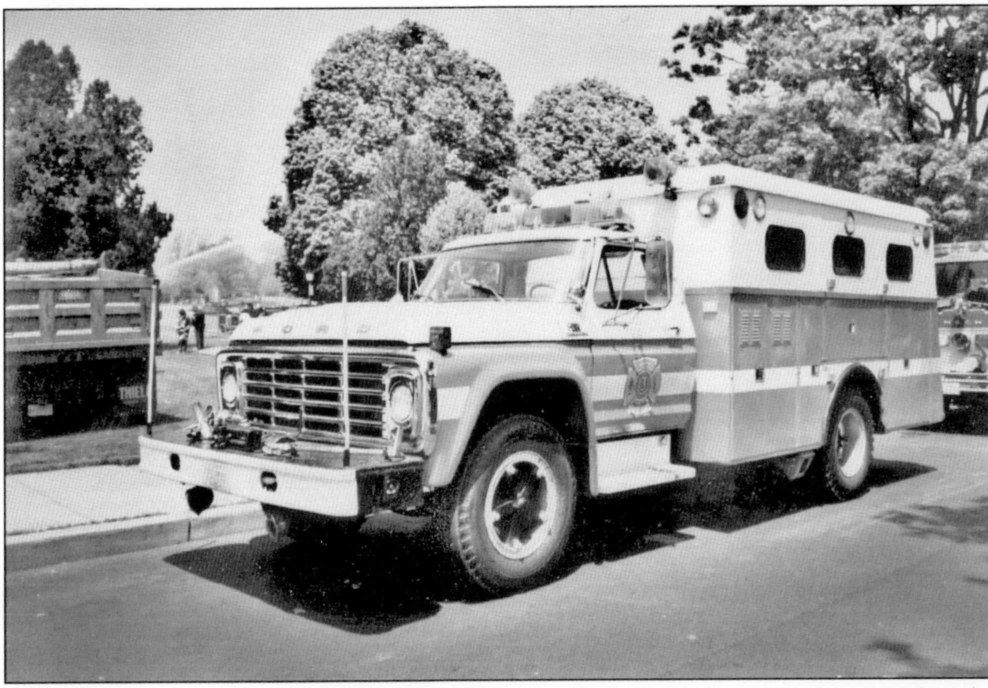

The 1976 Ford Squad, which replaced the Young as the primary rescue unit stationed at the Uniteds, was sold to the Blue Ridge Mountain Fire Company in West Virginia. (Courtesy Warren Jenkins.)

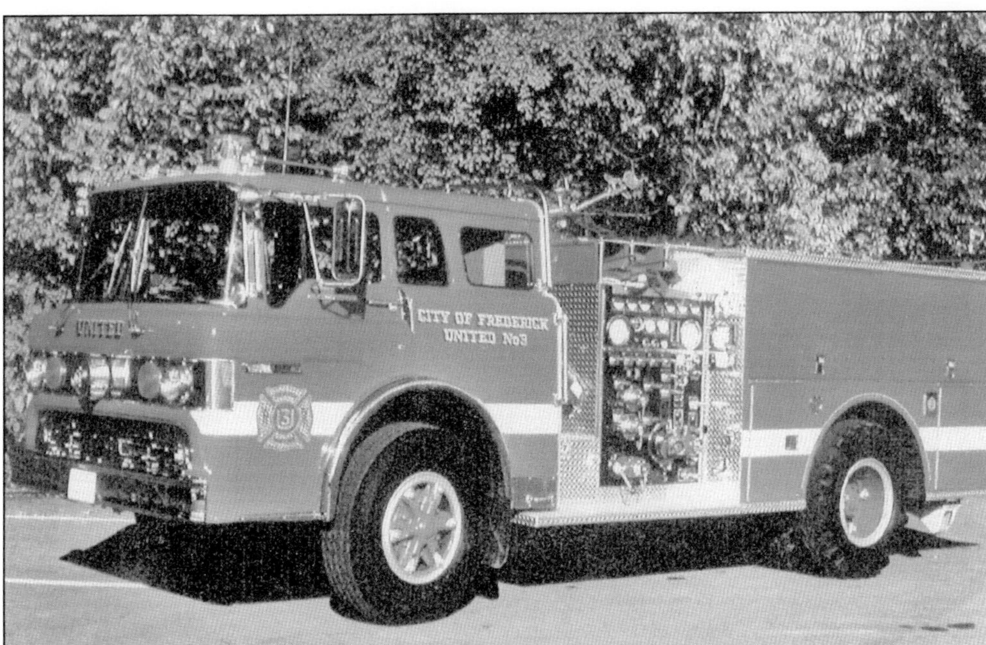

The Uniteds received the 1980 Ford Pierce from the City of Frederick and retired the Ahrens-Fox. (Courtesy Warren Jenkins.)

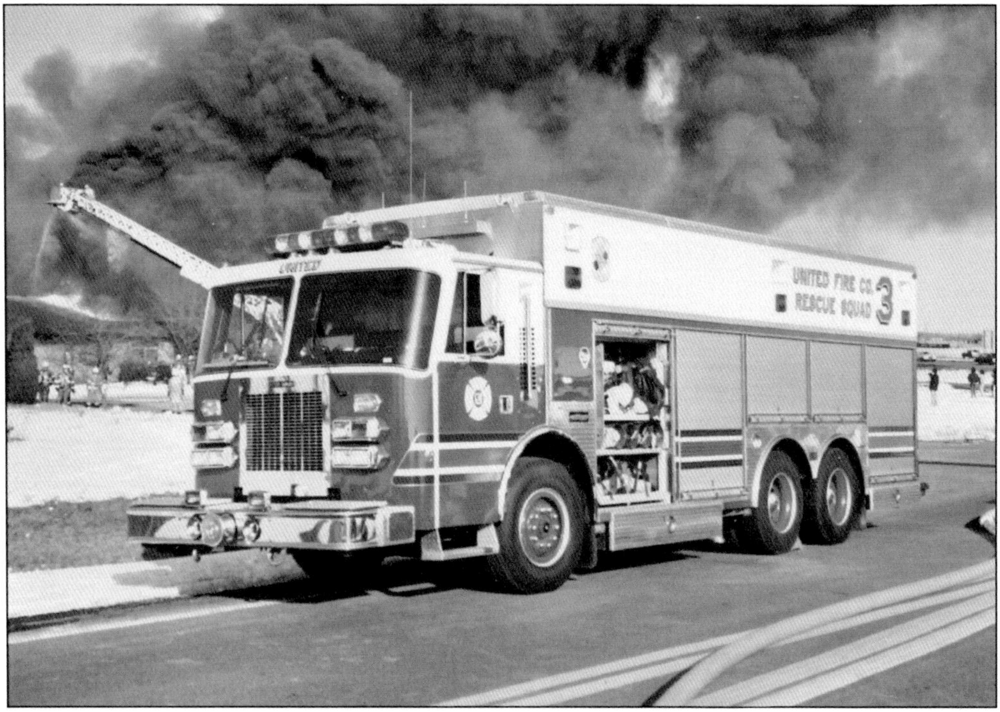

Squad 3, a 1991 Simon Duplex, is seen in action at the Seventh Day Adventist fire on February 28, 1993. (Courtesy Warren Jenkins.)

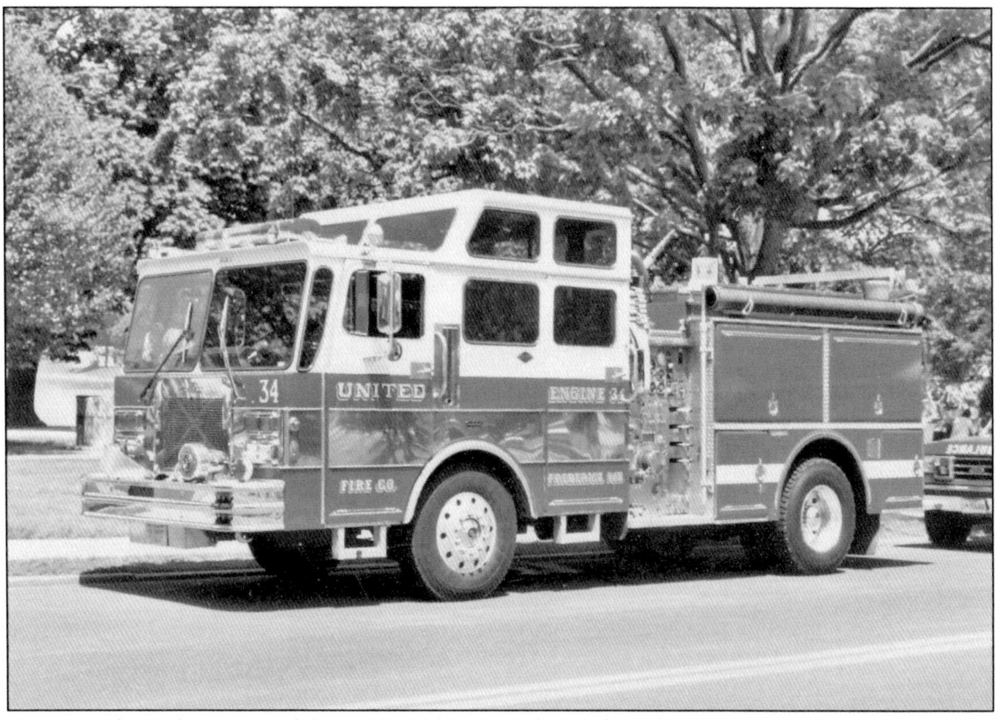

To respond to other areas of the county, the Uniteds purchased a 1985 Grumman pumper with a high headroom cab. (Courtesy Warren Jenkins.)

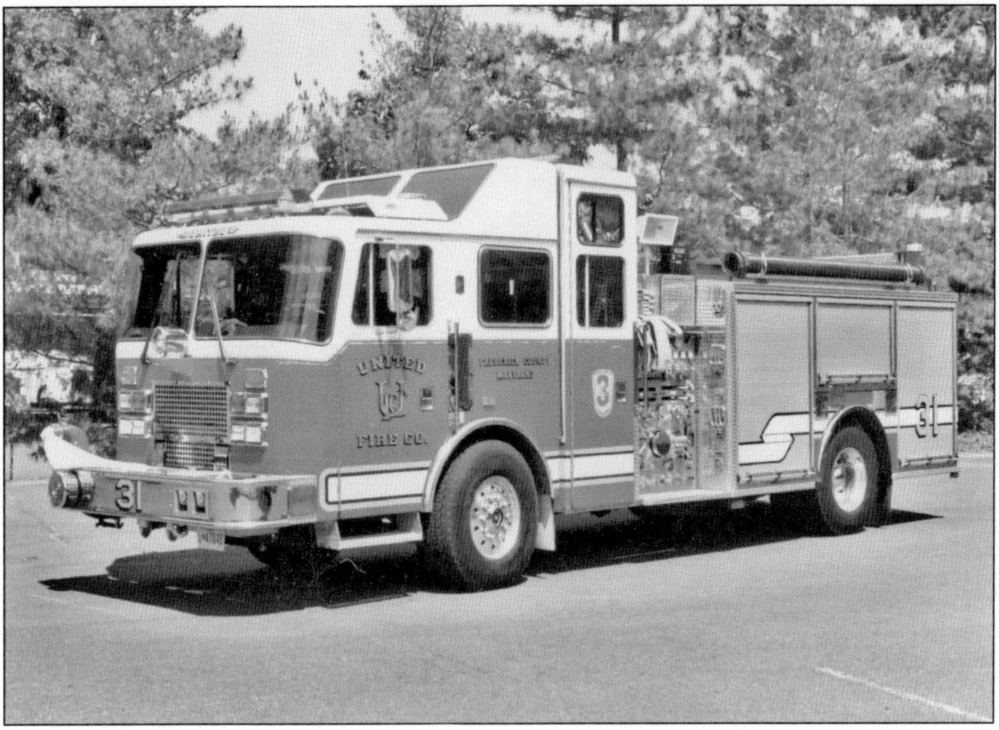

Engine 311 (formerly Engine 31), a 1996 KME, was originally assigned to Station 3 but moved to Station 31, the Westview Station, upon Westview's opening. (Courtesy Warren Jenkins.)

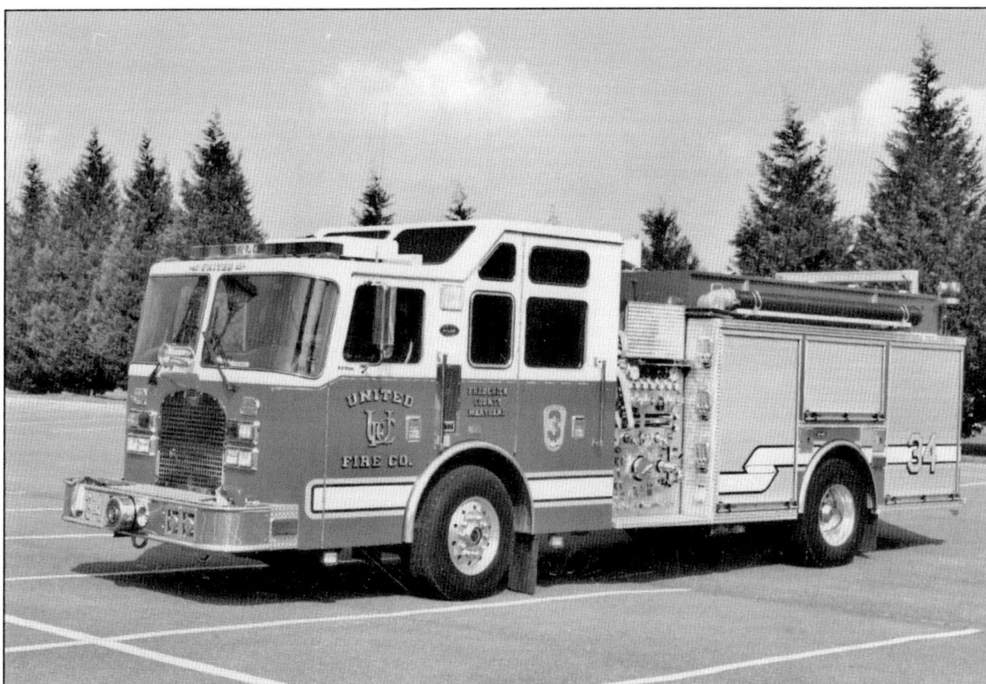

Engine/Tanker 34 is a 1999 KME pumper that responds from Station 3 on South Market Street. (Courtesy Warren Jenkins.)

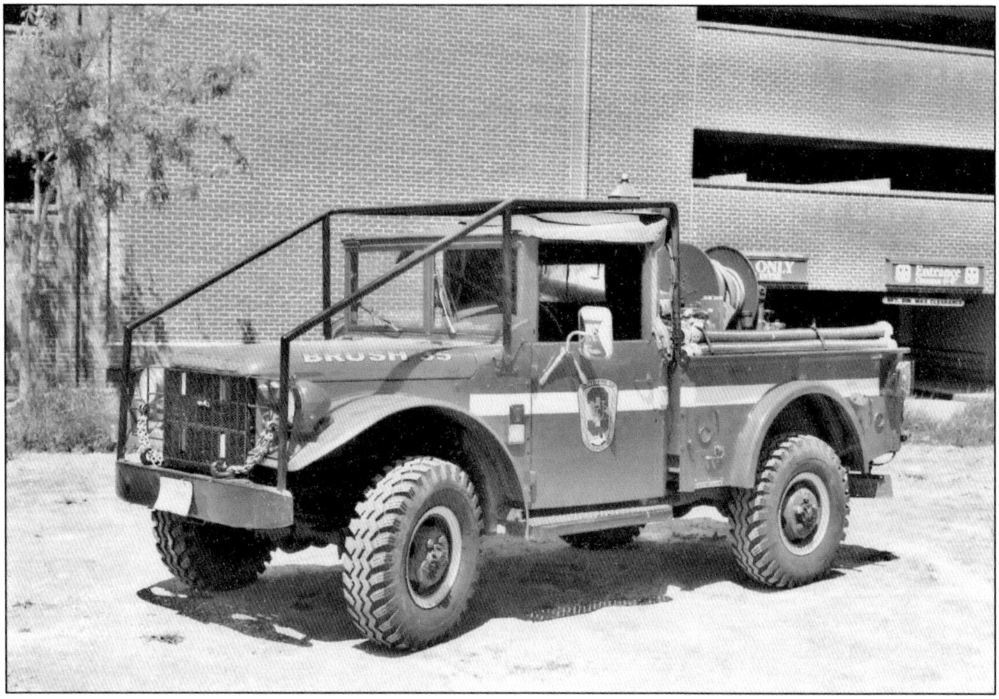

Brush 35 is a 1961 Dodge Power Wagon used to respond to grass and weeds fires around the Frederick area. (Courtesy Warren Jenkins.)

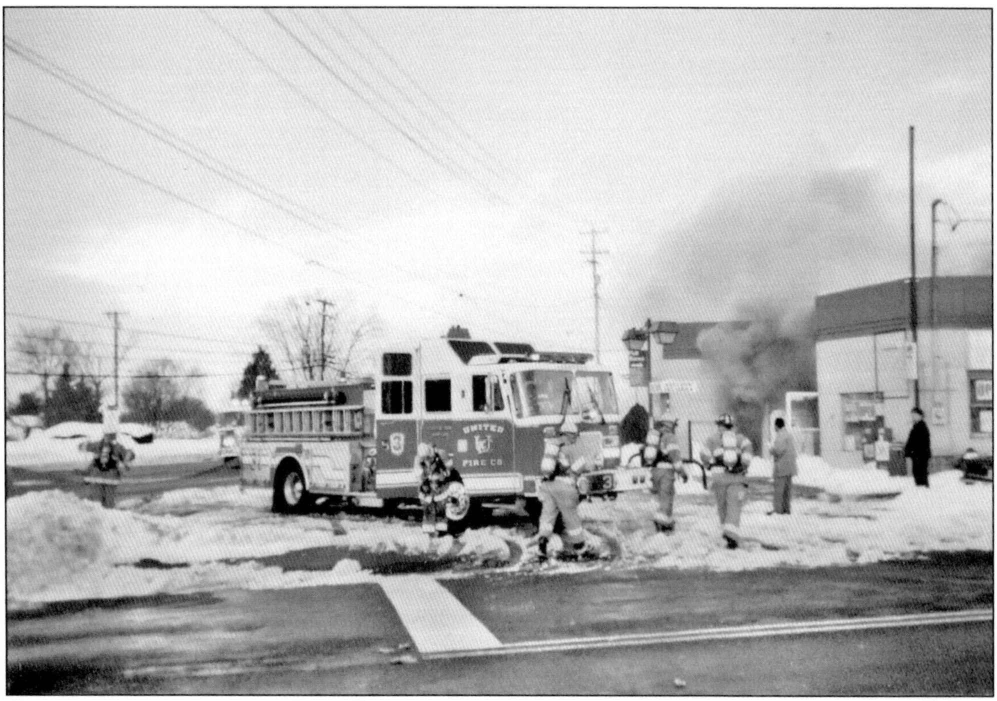

Engine 31, prior to its transfer to the Westview Station, is seen in action at a gas station fire on South Jefferson Street in Frederick.

Four

Citizens Truck Company No. 4

Company A

Though ladder service had been provided by the Junior Fire Company since 1878, the engine companies banded together with local utility companies to form a separate fire company with the goal of purchasing a motorized ladder truck to service Frederick. In 1926, the Citizens Truck Company was formed.

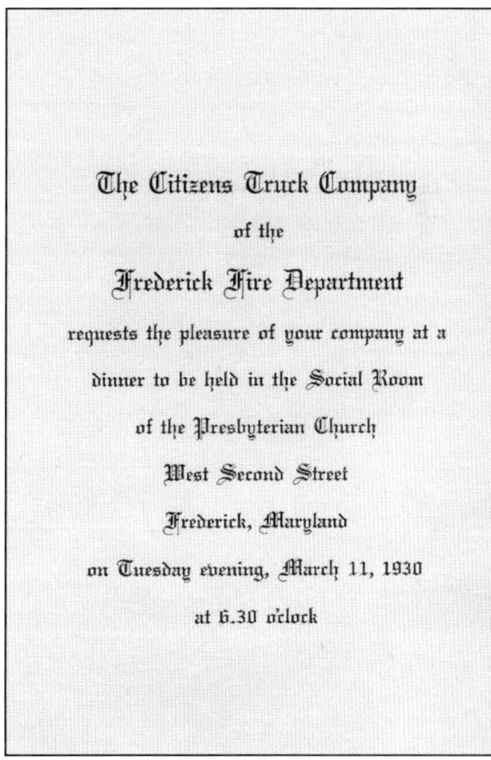

The Citizens Truck Company
of the
Frederick Fire Department
requests the pleasure of your company at a
dinner to be held in the Social Room
of the Presbyterian Church
West Second Street
Frederick, Maryland
on Tuesday evening, March 11, 1930
at 6.30 o'clock

Here is an invitation to a 1930 dinner of the Citizens Truck Company. Note that though there was never a formal Frederick Fire Department, the invitation indicates FFD existed. (Courtesy Sgt. Tom Lawson.)

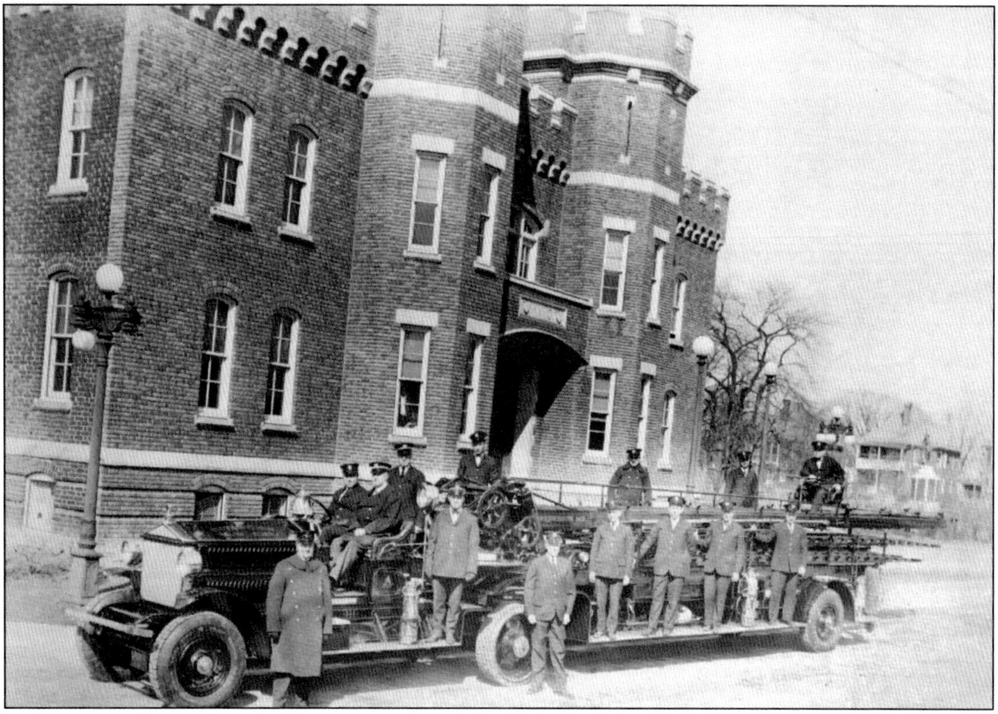

The first aerial apparatus of the Citizens, a 1927 Aherns-Fox, is displayed in front of the armory with many members aboard. (Courtesy UFC.)

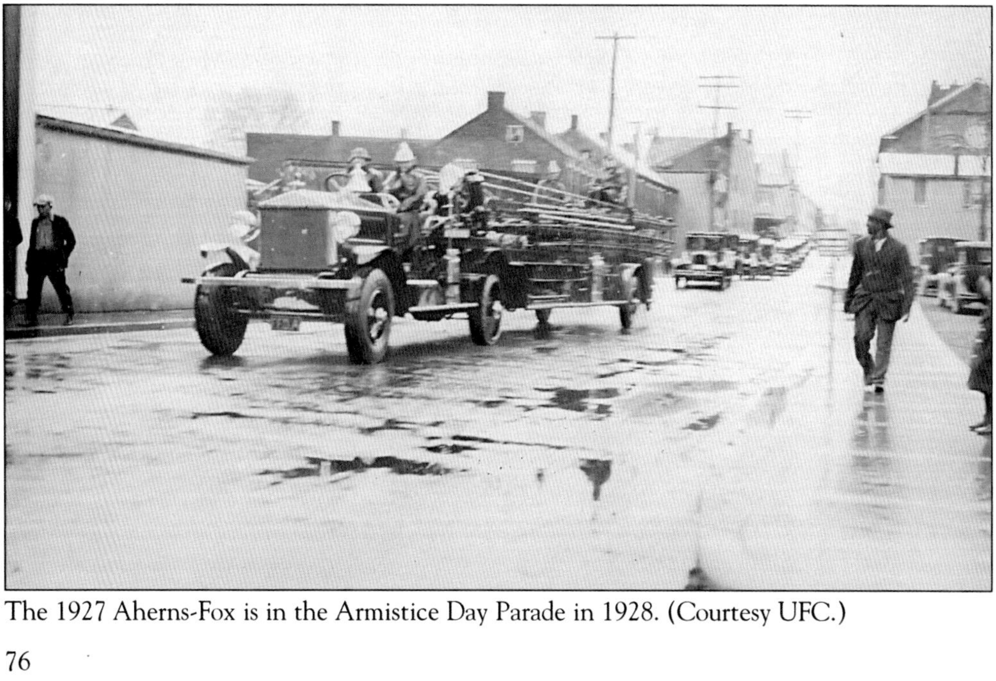

The 1927 Aherns-Fox is in the Armistice Day Parade in 1928. (Courtesy UFC.)

Training on March 3, 1929, included jumping into a life net. At the top of the aerial is Francis Davis of the Junior Fire Company, preparing to leap. (Courtesy UFC.)

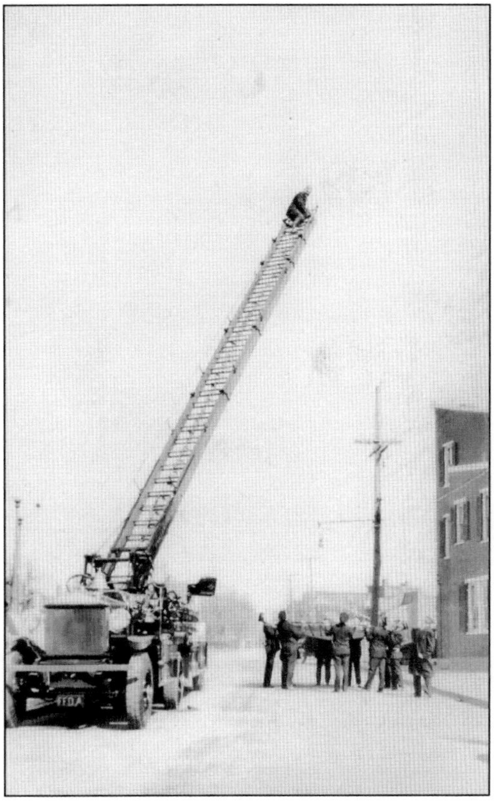

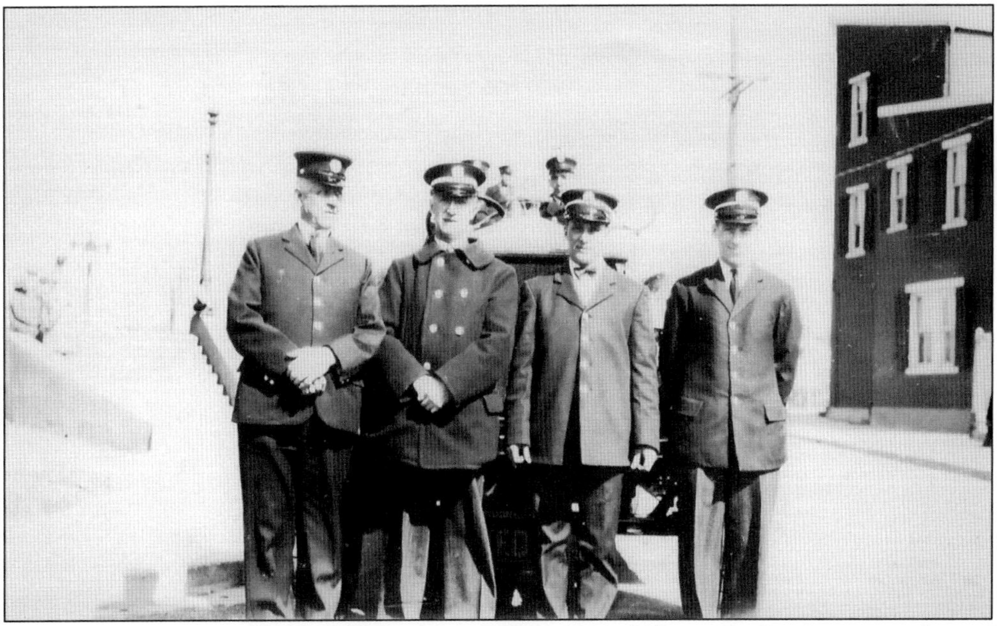

Among those in this photo taken at the March 3, 1929 drill are Dudley Page, president of the Citizens Truck Company, and F. Lester Smith, captain. Captain Smith was later elected president of the Maryland State Firemen's Association. (Courtesy UFC.)

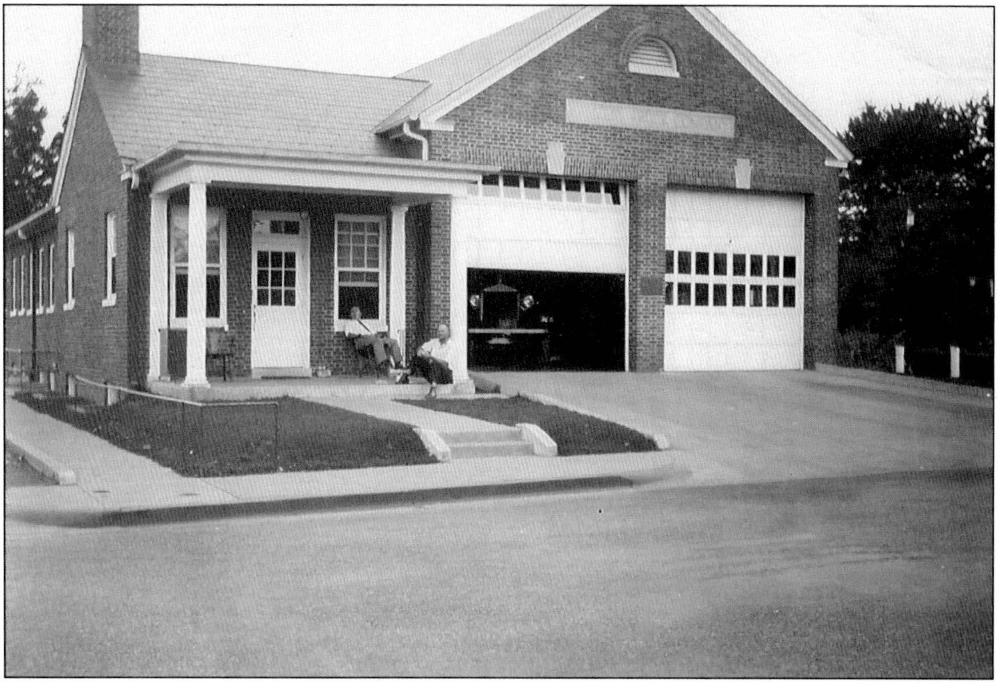

Built in 1939, the quarters of the Citizens Truck Company are on South Court Street. Note the Aherns-Fox ladder truck behind the door. (Courtesy H. Richard Hahn.)

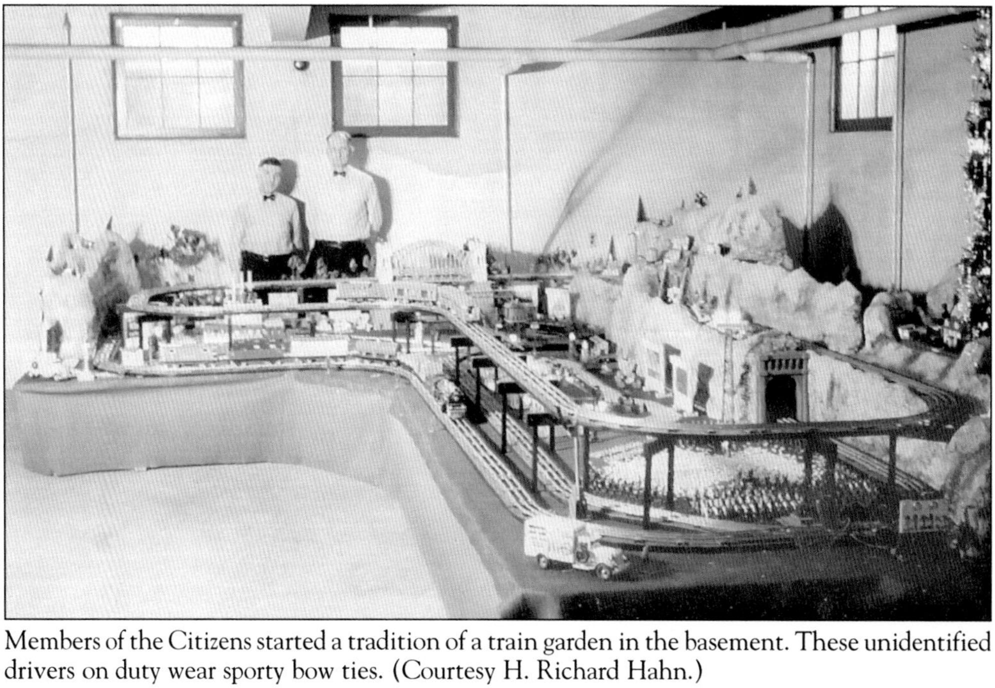

Members of the Citizens started a tradition of a train garden in the basement. These unidentified drivers on duty wear sporty bow ties. (Courtesy H. Richard Hahn.)

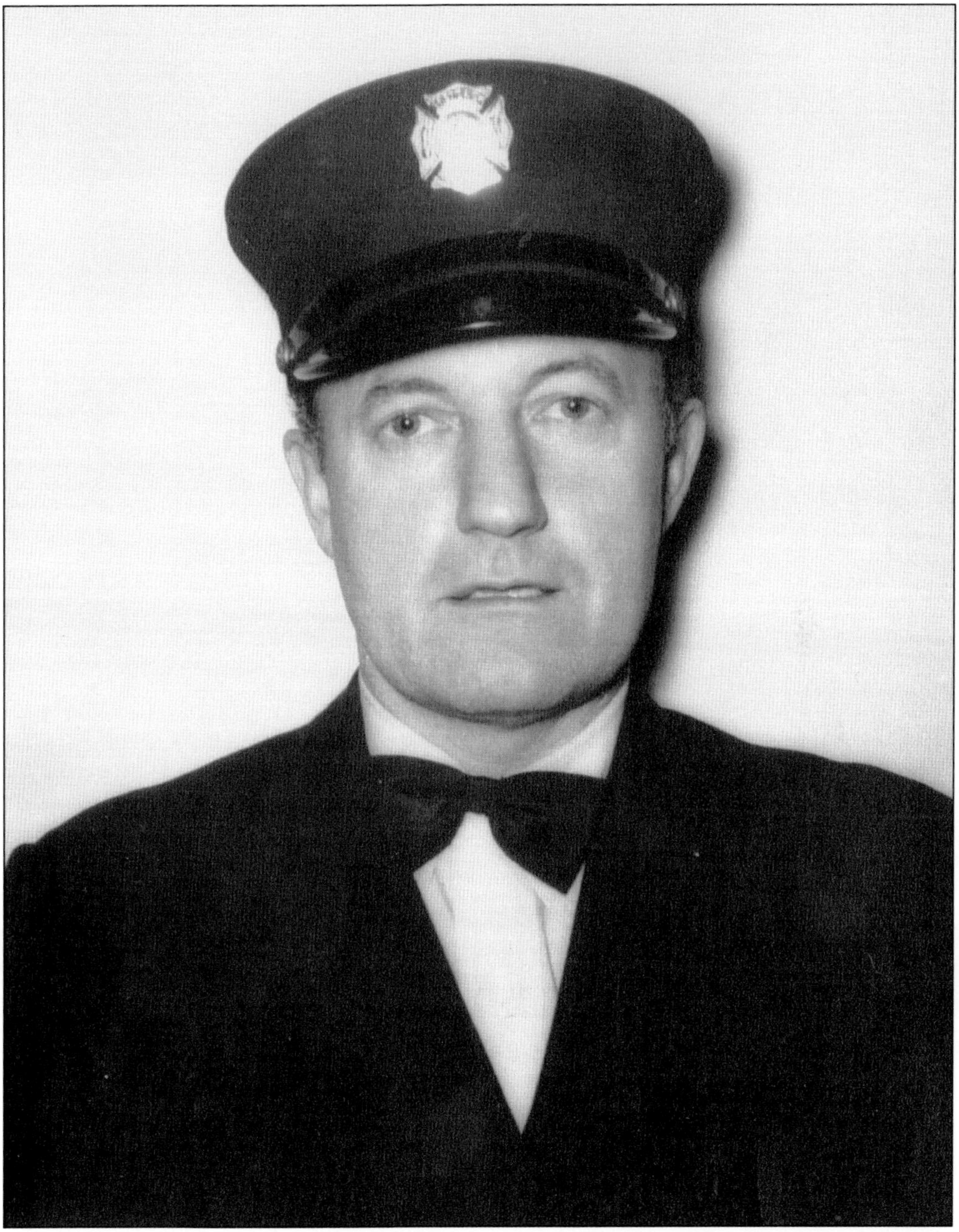

Ray Steele served as a driver of the Citizens Truck Company for 35 years. He was credited with saving at least 12 lives during his time as a firefighter. He was also well known as a player/manager of the Frederick Hustlers semi-pro baseball team. (Courtesy Sgt. Thomas Lawson.)

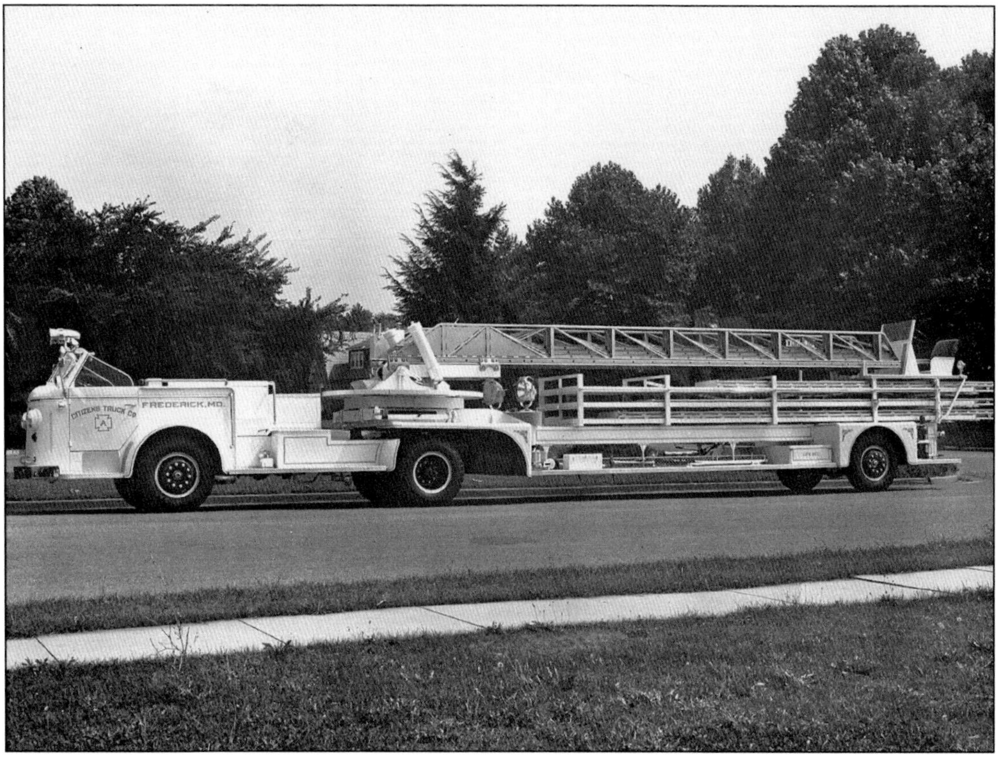

The sleek 1951 American LaFrance aerial ladder was a tremendous addition to the city fire protection. This unit was state of the art when delivered. Note the door shows Company A in the Maltese cross. (Courtesy JFC.)

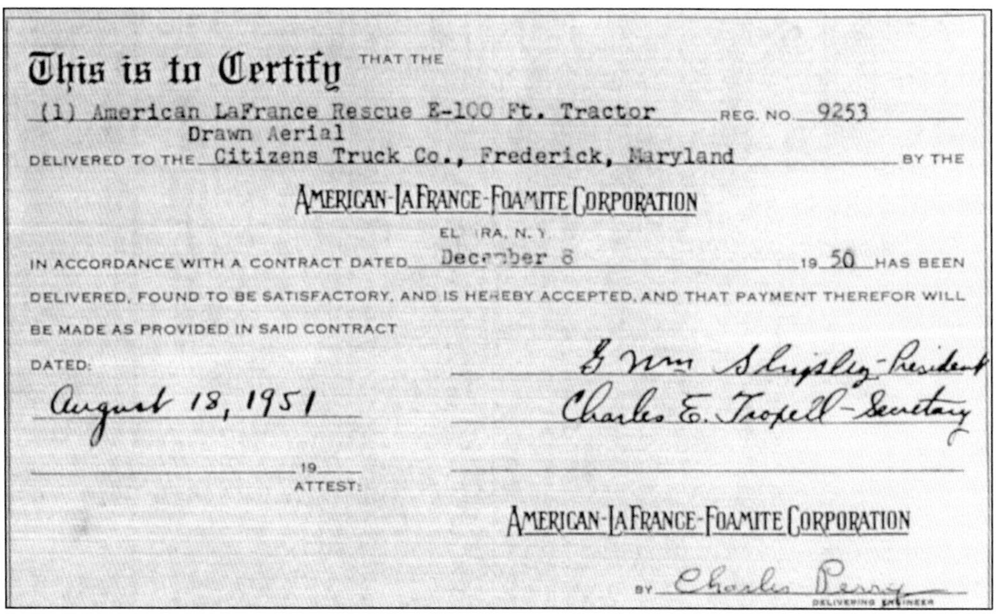

This is to Certify THAT THE

(1) American LaFrance Rescue E-100 Ft. Tractor Drawn Aerial REG. NO. 9253

DELIVERED TO THE Citizens Truck Co., Frederick, Maryland BY THE

AMERICAN-LAFRANCE-FOAMITE CORPORATION

EL IRA, N. Y

IN ACCORDANCE WITH A CONTRACT DATED December 8 19 50 HAS BEEN DELIVERED, FOUND TO BE SATISFACTORY, AND IS HEREBY ACCEPTED, AND THAT PAYMENT THEREFOR WILL BE MADE AS PROVIDED IN SAID CONTRACT

DATED:

August 18, 1951

G. Wm Shipley - President

Charles E. Troxell - Secretary

19

ATTEST:

AMERICAN-LAFRANCE-FOAMITE CORPORATION

BY Charles Perry

DELIVERING ENGINEER

Before being placed in service, the Citizens Truck Company received this original certification from the American LaFrance-Foamite Corporation of Elmira, New York. (Courtesy Sgt. Thomas Lawson.)

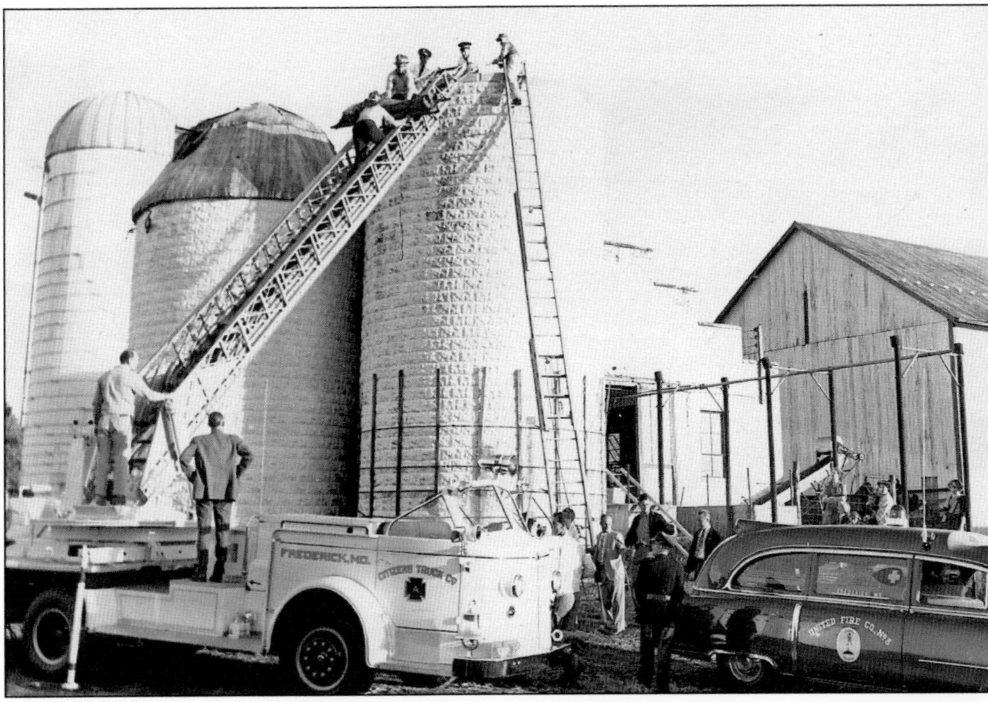

The ladder truck from Citizens was called to assist the United ambulance on a silo rescue near Frederick. (Courtesy Sgt. Thomas Lawson.)

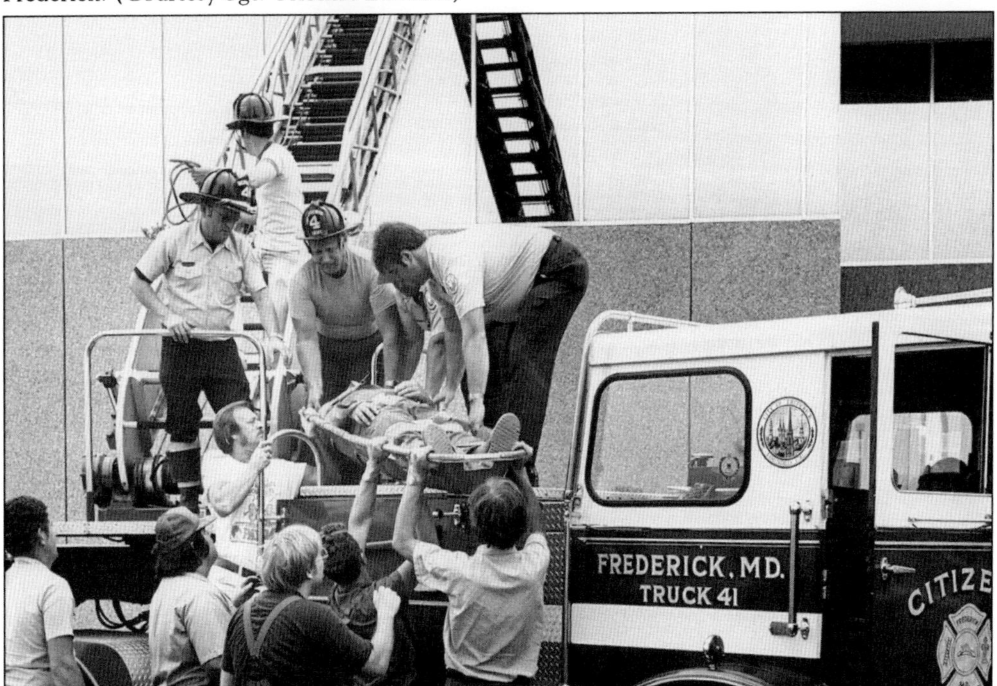

Another rescue required the Citizens ladder to remove an injured worker from the roof of the Aircraft Owners and Pilots Association (AOPA) building at the Frederick Airport. (Courtesy Sgt. Thomas Lawson.)

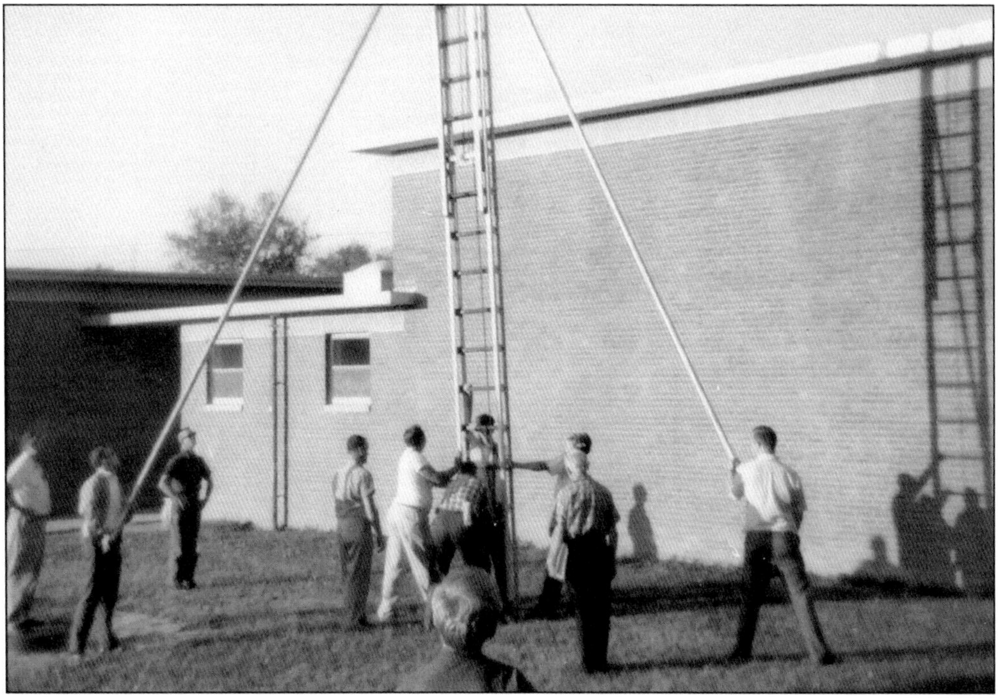

A drill in 1969 at the North Frederick School included the use of the "pole" or "Bangor" ladder. (Courtesy H. Richard Hahn.)

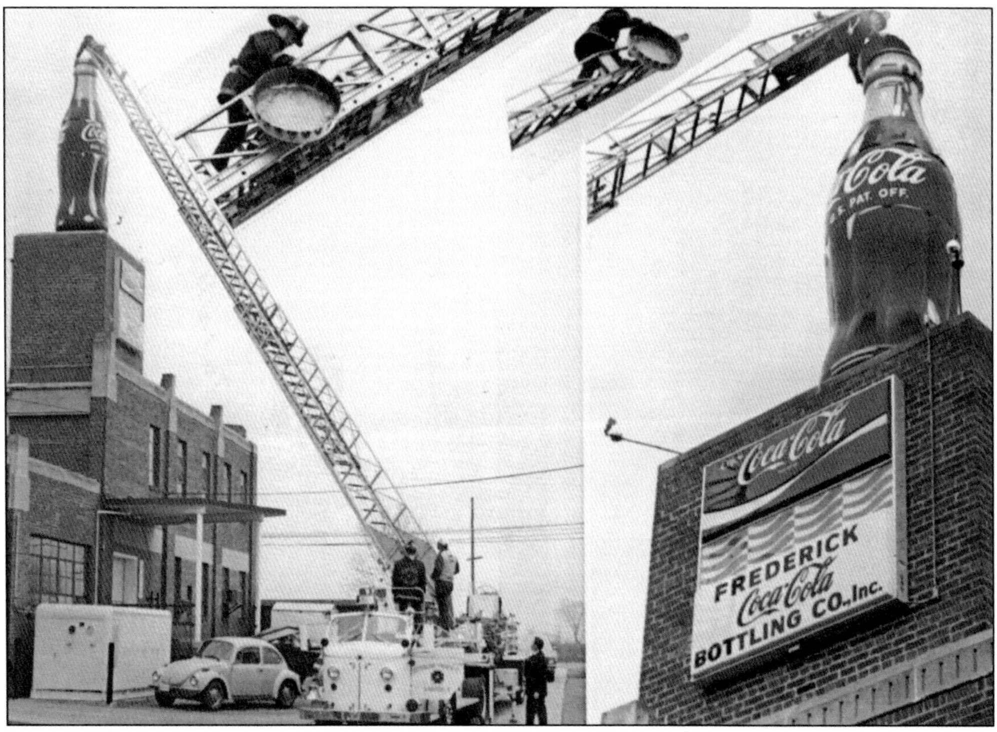

The Frederick Coca-Cola Bottling Co., Inc., could not cap this Coke bottle without the aid of Ladder 4. The top blew off during a severe windstorm. (Courtesy Sgt. Thomas Lawson.)

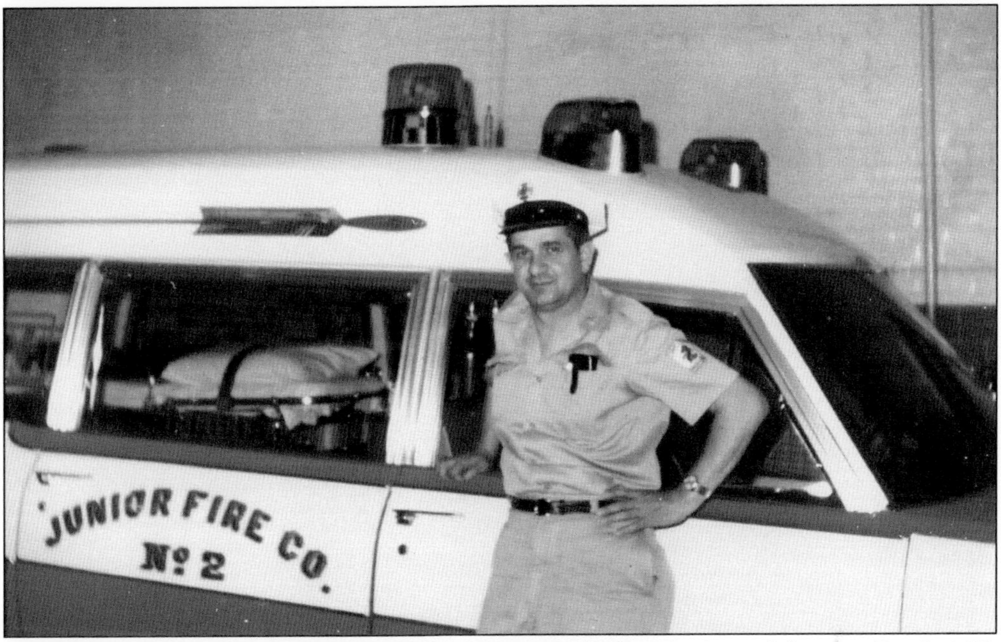

Capt. Barney Stroup of the Citizens Truck Company died in the line of duty on October 27, 1968, when he collapsed while returning from a false alarm. Also active with the Juniors, Captain Stroup is shown here as a crew member on the Junior ambulance. (Courtesy Sgt. Thomas Lawson.)

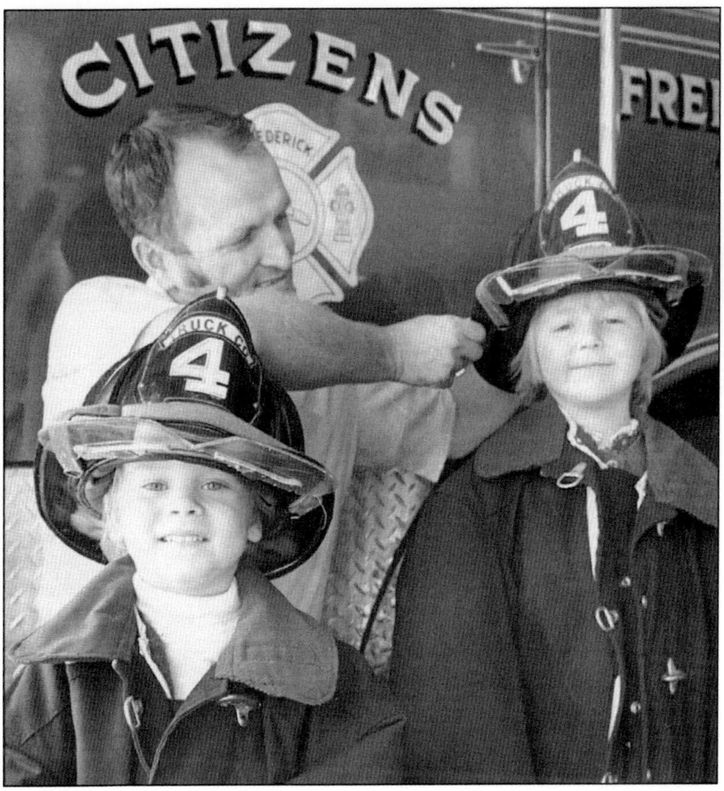

Children are always fascinated with the fire company. Sgt. Thomas Lawson brings smiles to a visiting school group from the Frederick Alliance Church Pre-School. (Courtesy Sgt. Thomas Lawson.)

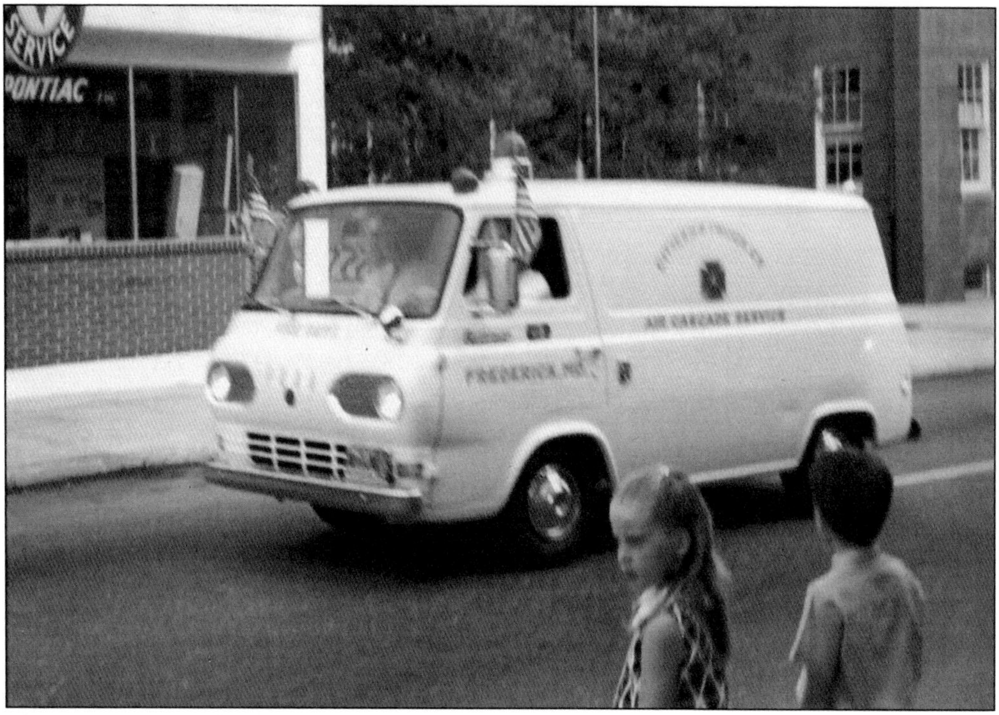

The Citizens were the first fire company in Frederick County to purchase a unit to fill self-contained breathing apparatus. Cascade 4, a 1967 Ford Econoline, is shown here in a Frederick parade. (Courtesy H. Richard Hahn.)

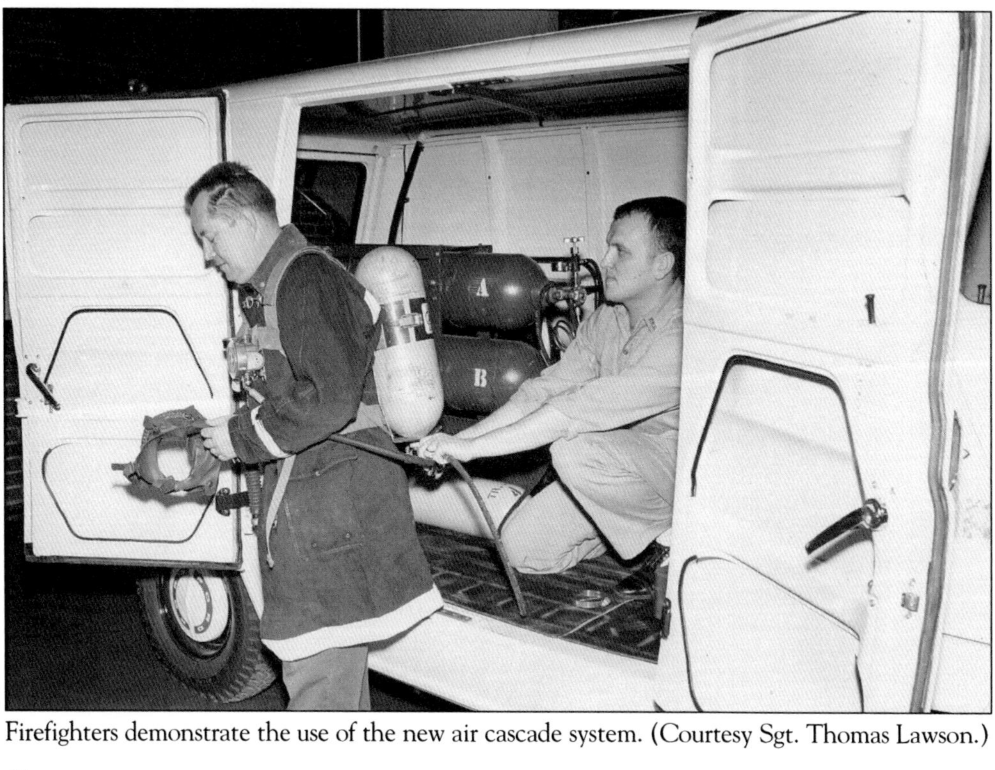

Firefighters demonstrate the use of the new air cascade system. (Courtesy Sgt. Thomas Lawson.)

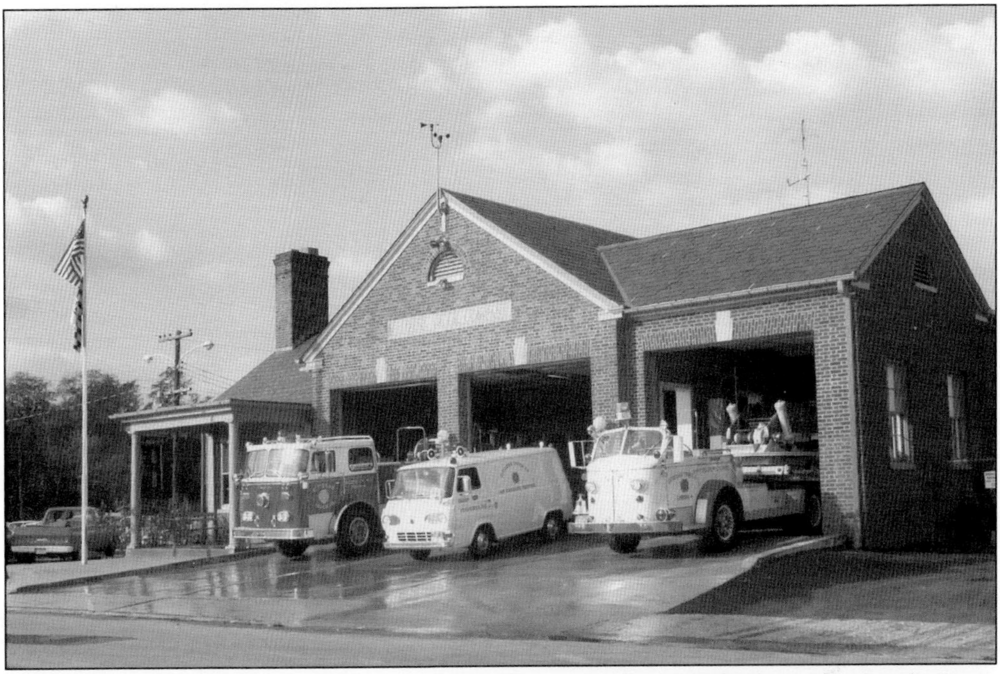

When the second ladder truck was purchased, an addition was built to the South Court Street station to accommodate both trucks and the air cascade unit. (Courtesy Sgt. Thomas Lawson.)

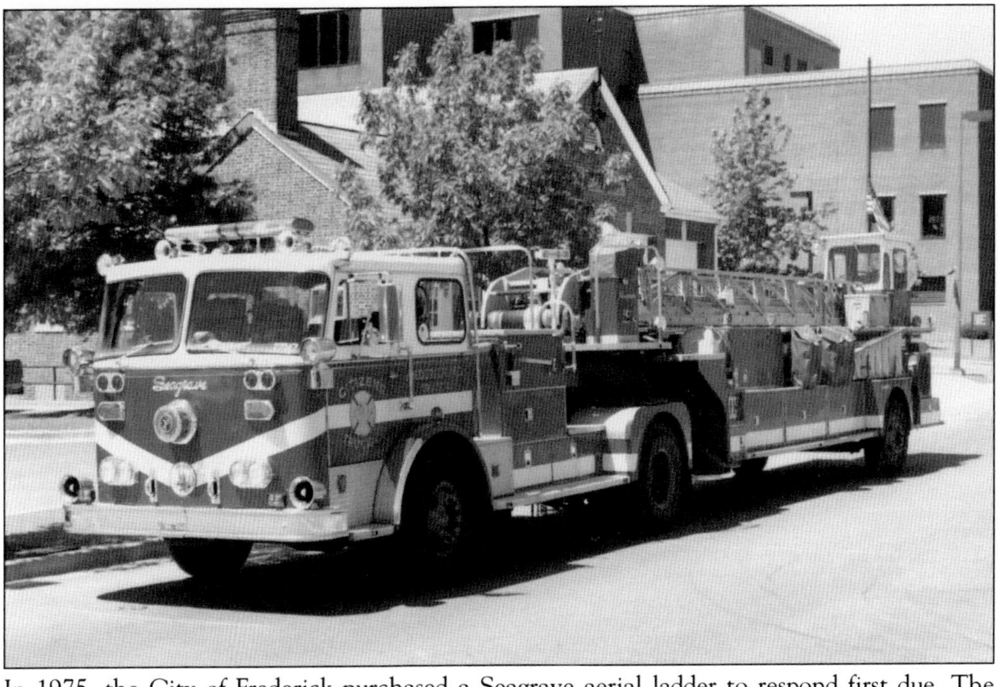

In 1975, the City of Frederick purchased a Seagrave aerial ladder to respond first due. The Citizens maintained the 1951 American LaFrance as a second due truck. At this time, the numbers changed from Ladder 4 to Truck 41 and Truck 42 respectively. (Courtesy Warren Jenkins.)

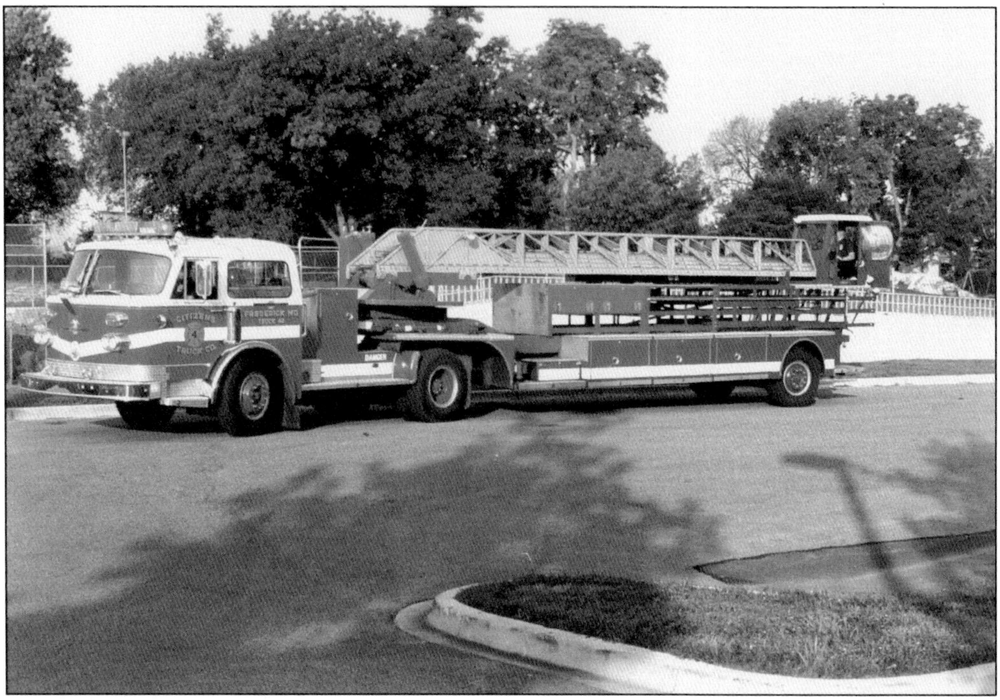

The 1951 American LaFrance was reconditioned and re-powered in the 1980s with a new American LaFrance tractor and fully enclosed tiller position. (Courtesy Warren Jenkins.)

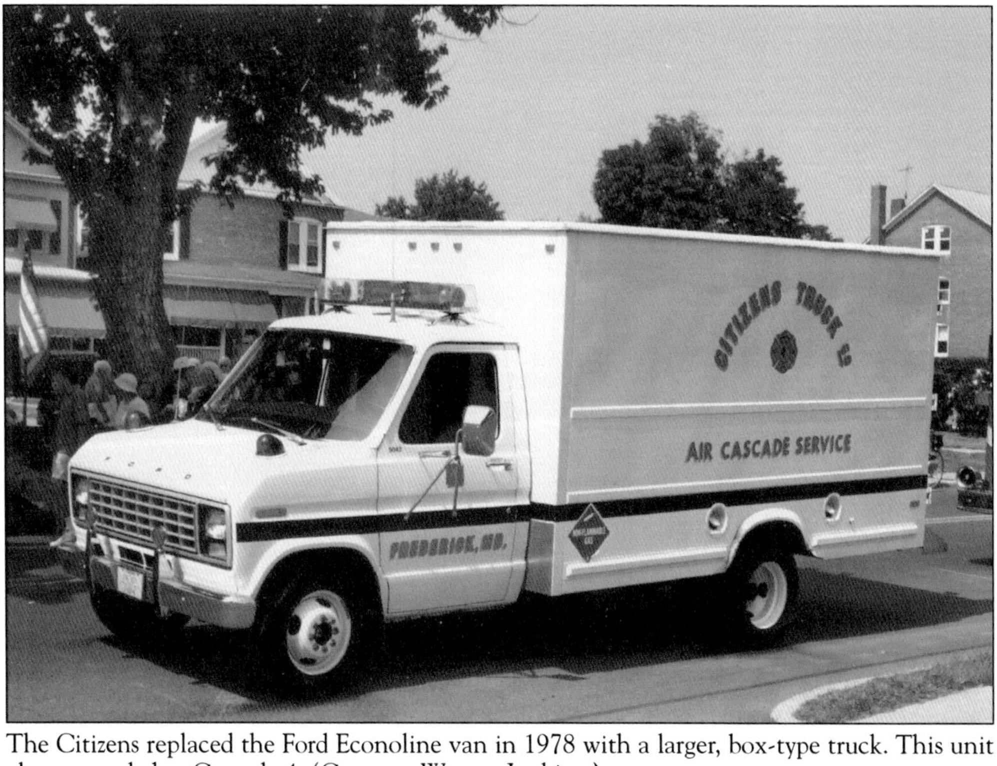

The Citizens replaced the Ford Econoline van in 1978 with a larger, box-type truck. This unit also responded as Cascade 4. (Courtesy Warren Jenkins.)

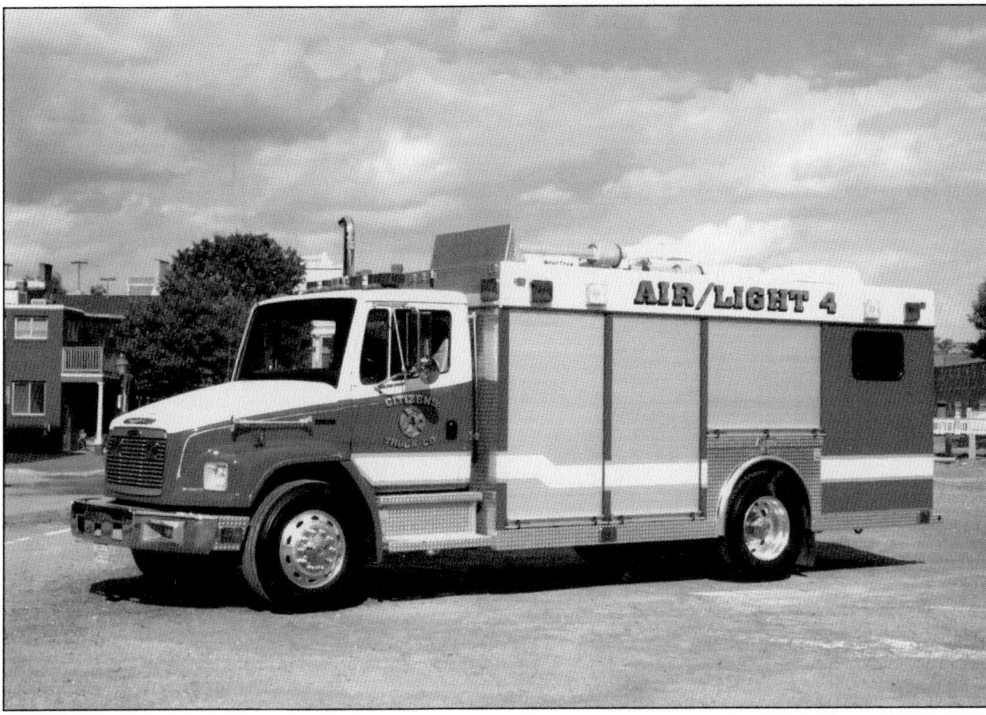

Air Unit 4 is a 1998 Freightliner Summit. The latest air unit is also equipped as a light unit. (Courtesy Warren Jenkins.)

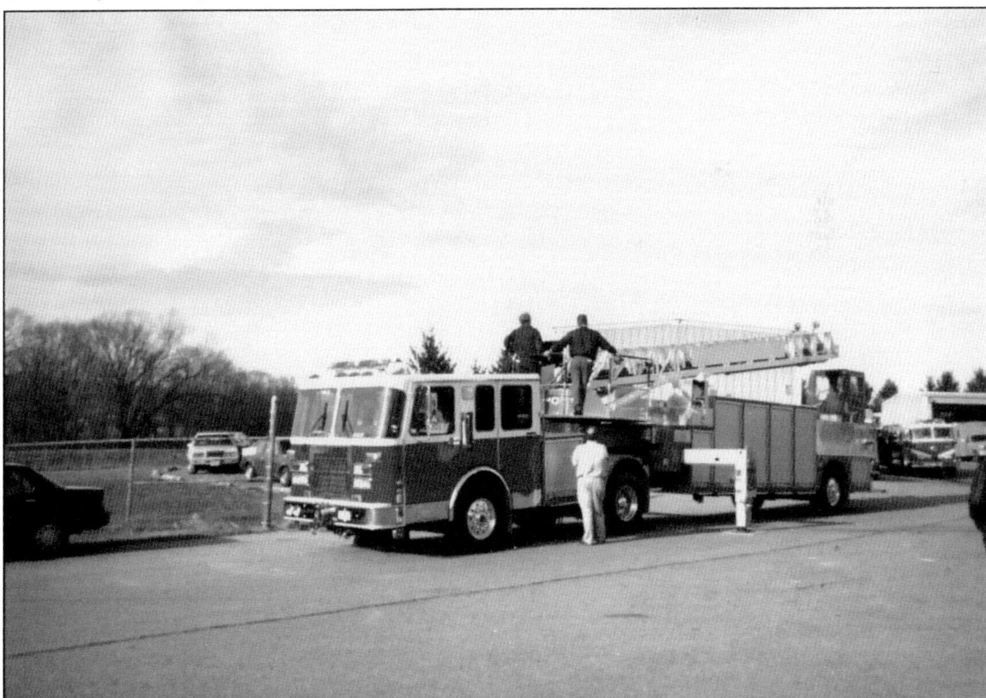

Truck 41 is a 1994 100-foot Spartan LTI tractor-drawn aerial, shown here during the acceptance test. (Courtesy Warren Jenkins.)

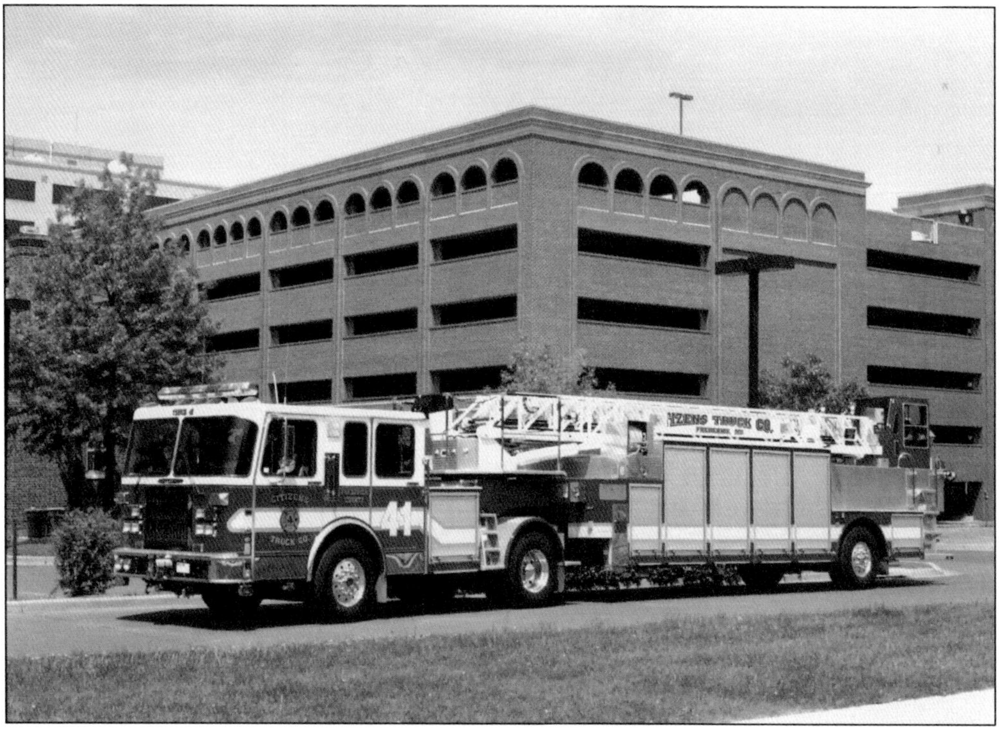

Truck 41 is shown after it was placed in service in front of the Court Street Parking Deck opposite the Citizens Truck Company quarters. (Courtesy Warren Jenkins.)

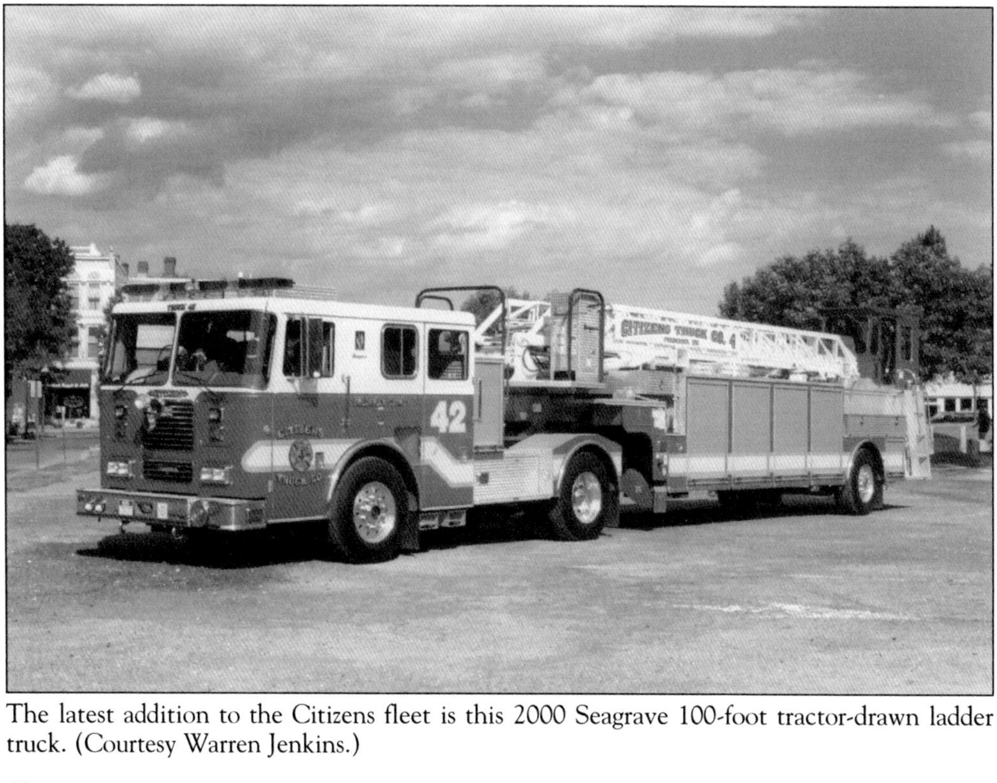

The latest addition to the Citizens fleet is this 2000 Seagrave 100-foot tractor-drawn ladder truck. (Courtesy Warren Jenkins.)

Five

WORKING FIRES

Though Frederick has always been proud of the excellent fire protection provided by the local fire service, major fires have occurred. However, it should be noted that all major fires have been contained to the building of origin. Following are some of the more memorable fires in Frederick's history.

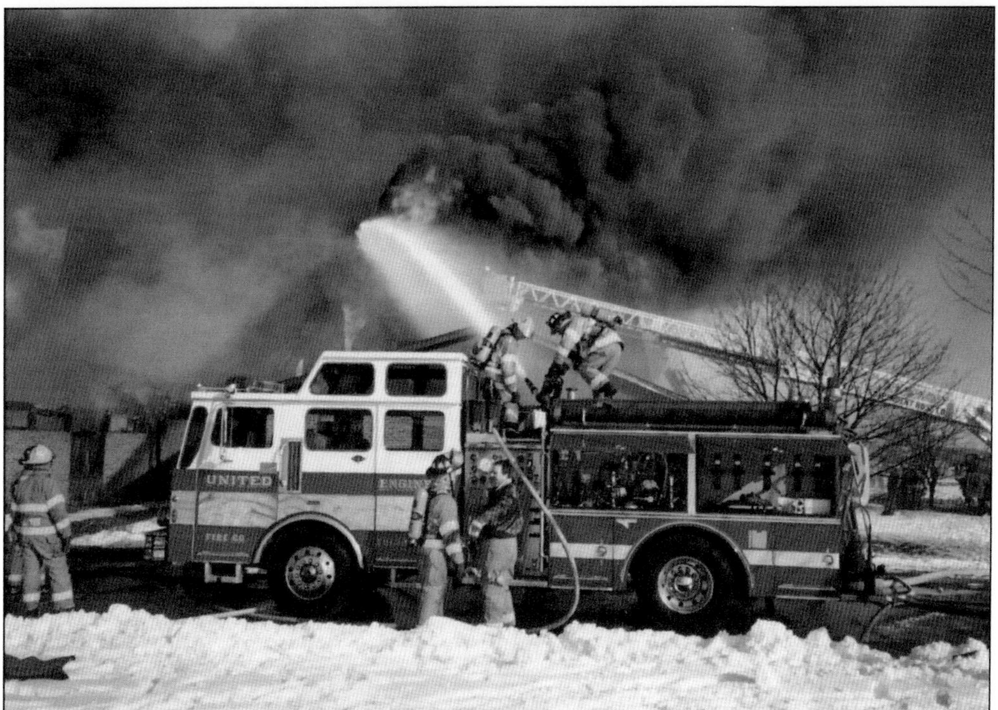

One of the largest fires in the 1990s was the Seventh Day Adventist Church fire on February 28, 1993. Here, Engine 33 and Tower 1 flow large volumes of water in an aggressive exterior attack. (Courtesy Warren Jenkins.)

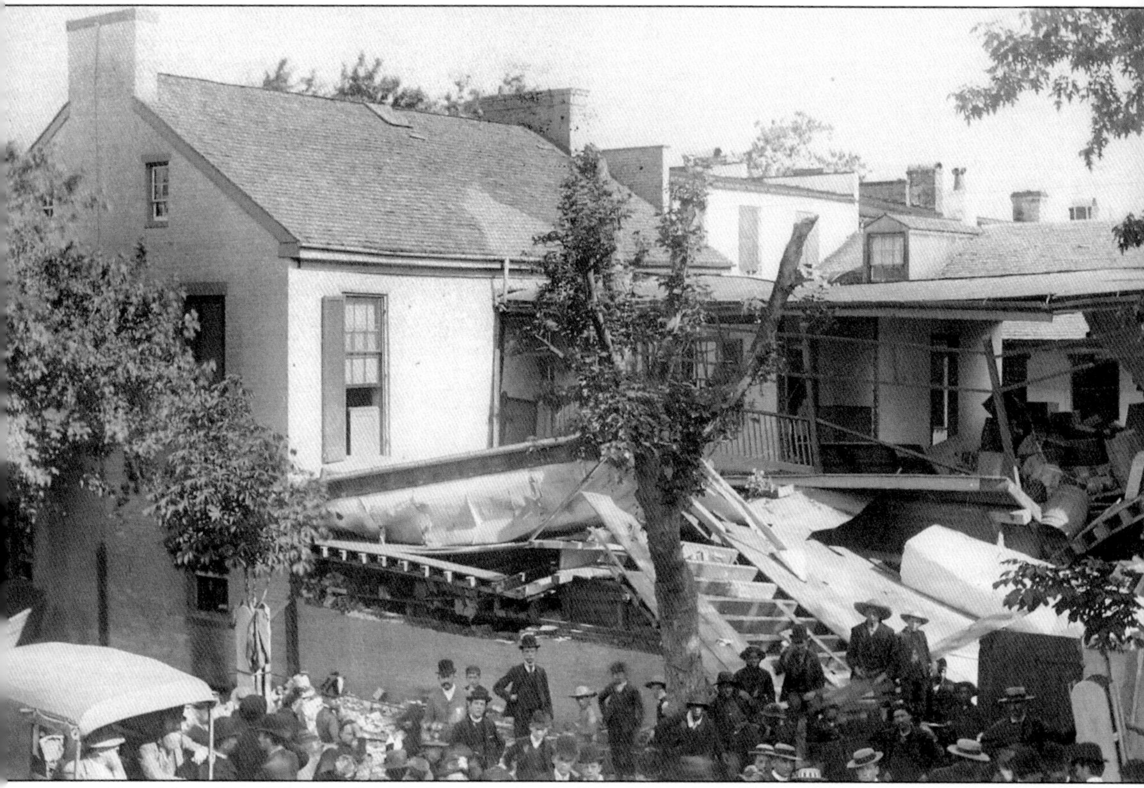

On May 28, 1888, a fire and massive explosion occurred at Zeller's Store at South and Market Street. A store clerk was killed and over 30 members of the United Fire Company injured. Though no line of duty deaths were recorded at the time, the records of the United Fire Company show an unusual number of deaths of members in the year following the fire, presumably from injuries and burns. (Courtesy UFC.)

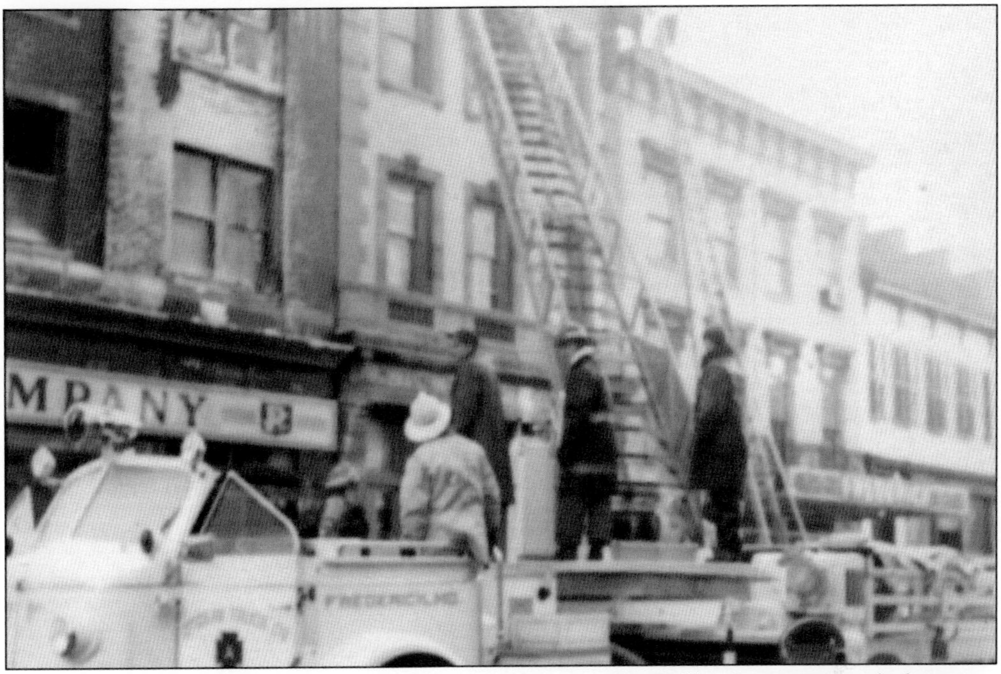

It was nine degrees above zero at 4:55 a.m. on January 31, 1955, when a general alarm was sounded for J.C. Penney Department Store at 114–118 North Market Street in downtown Frederick. The building was completely gutted by fire, but the hard work of firefighters from Frederick, Fort Detrick, and other neighboring communities kept this blaze from spreading. This fire would most likely be the largest property loss in Frederick history by today's pricing standards. (Courtesy Sgt. Thomas Lawson.)

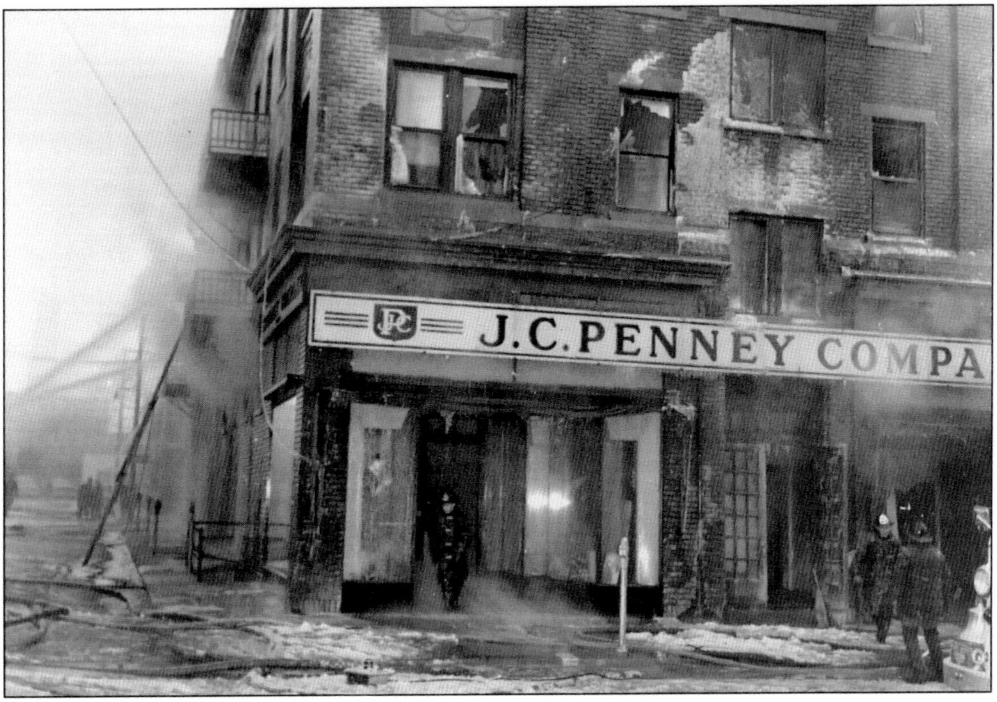

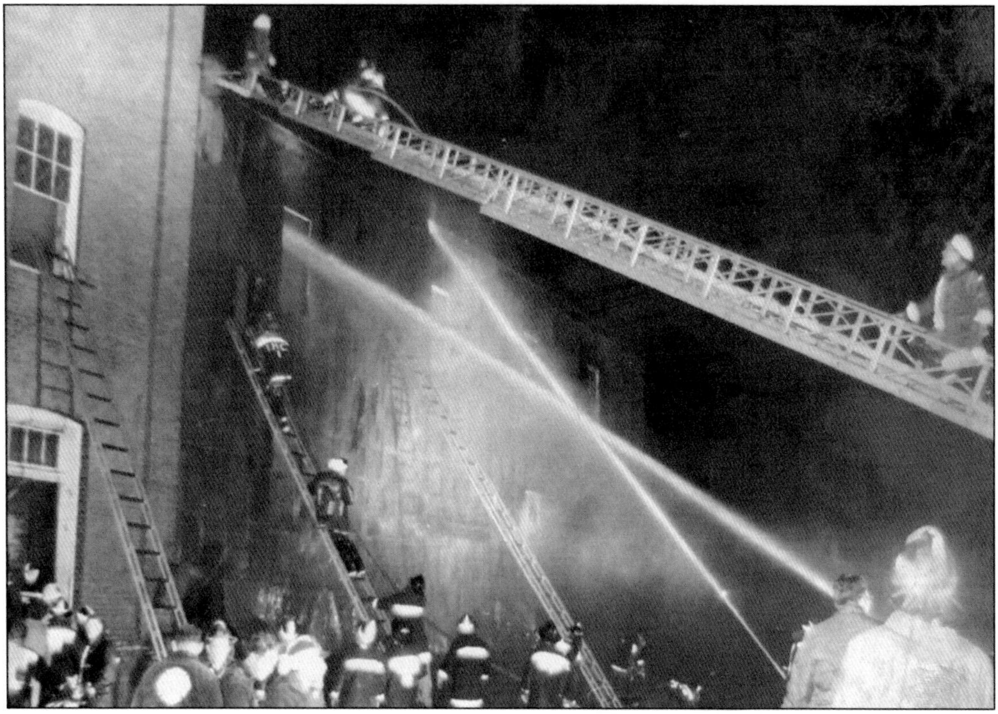

On April 9, 1968, the city saw civil unrest after the assassination of Dr. Martin Luther King Jr. A large fire broke out in the former Mountain City Mill on South Carroll Street. This was the first of several major fires in that structure. (Courtesy Sgt. Thomas Lawson.)

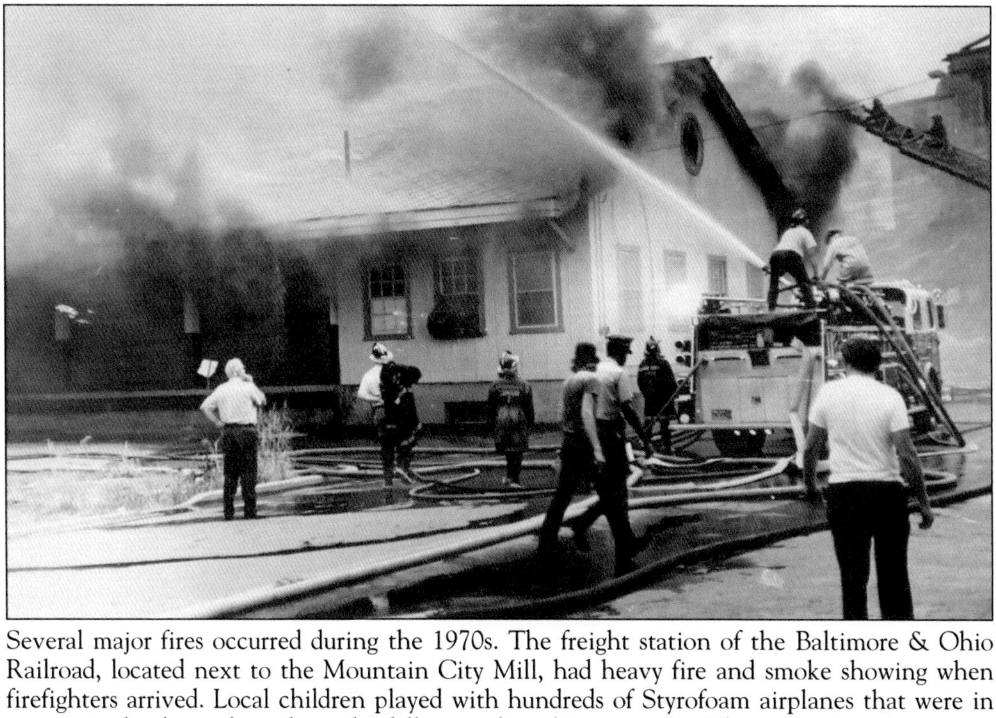

Several major fires occurred during the 1970s. The freight station of the Baltimore & Ohio Railroad, located next to the Mountain City Mill, had heavy fire and smoke showing when firefighters arrived. Local children played with hundreds of Styrofoam airplanes that were in storage in the depot throughout the following days. (Courtesy Sgt. Thomas Lawson.)

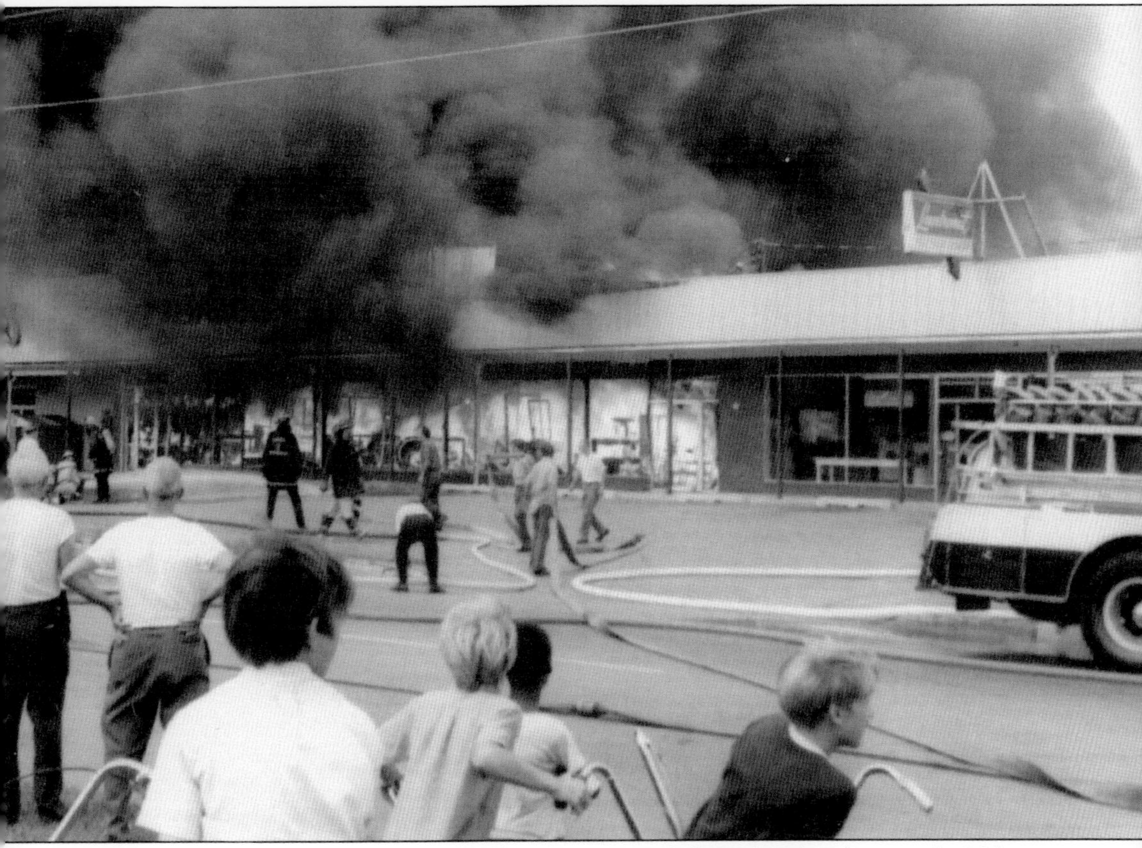

On a hot August day in 1971, the Frederick police responded to a burglar alarm at the Hiltner Furniture Store at 919 East Street. Unable to see into the large plate glass windows, police soon realized they were looking into a smoke bank and a fire was inside the building. Ultimately, several surrounding companies were required to assist the Frederick units. (Courtesy JFC.)

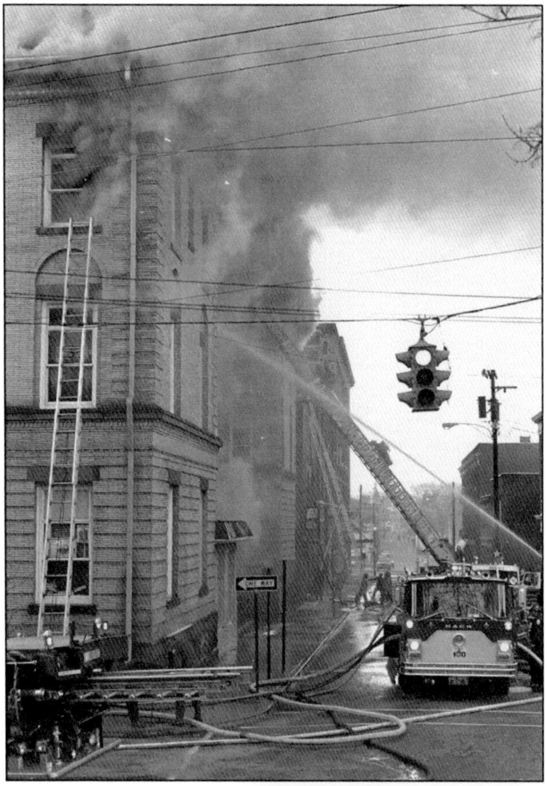

On December 7, 1974, a major fire destroyed the former YMCA located at the intersection of Church and Court in Frederick. Ladder trucks from both Hagerstown and Rockville responded, and a ladder from Kensington stood by at the Citizens. (Courtesy Sgt. Thomas Lawson.)

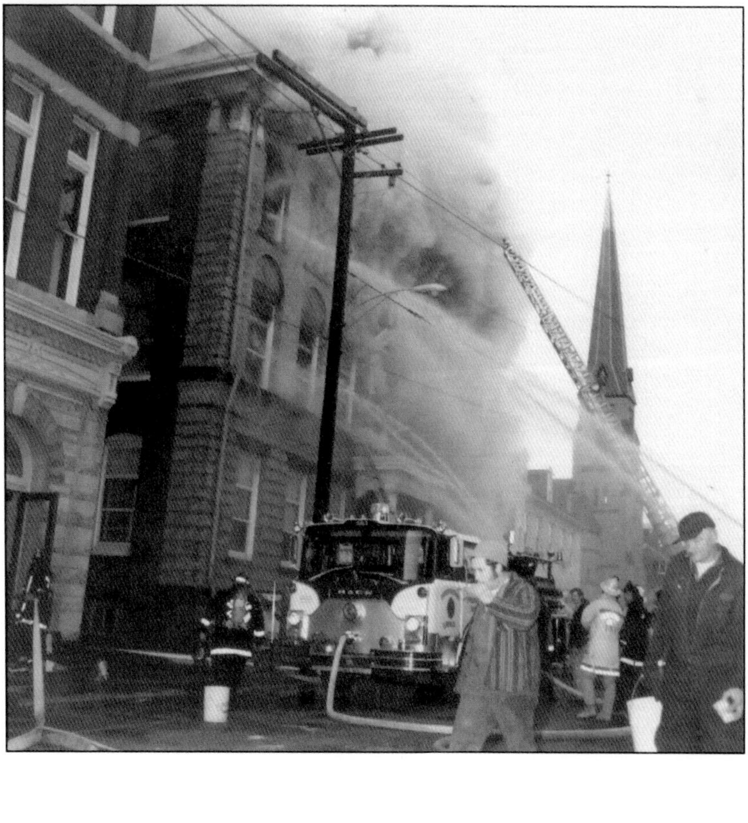

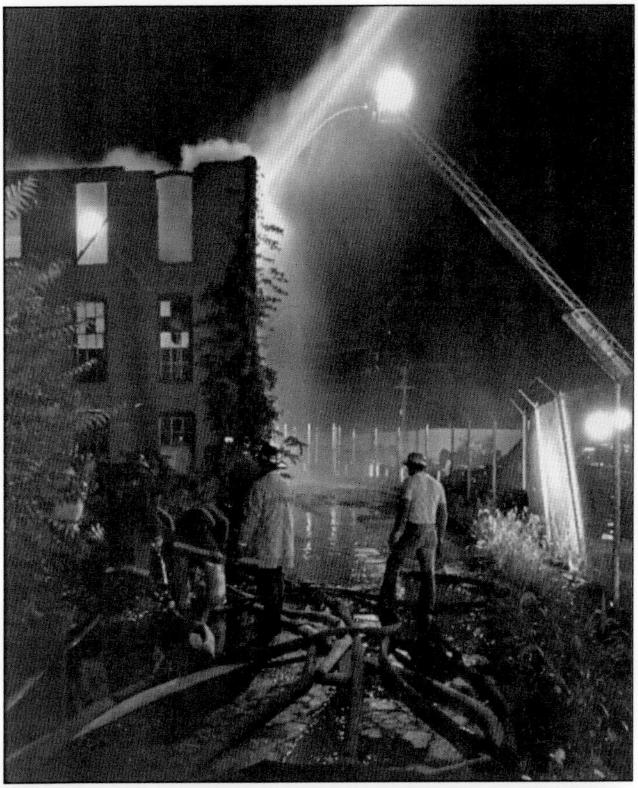

During a severe summer storm, lightning struck the water tower of the Goodwill complex (formerly the Ox Fiber Brush Company) on East Church Street in Frederick. Suddenly, fire began to show from the roof area, ultimately destroying almost a full acre of industrial buildings. In the top photo, an aerial from Halfway, Maryland, begins operations. The bottom photo shows the extent of the devastation. (Courtesy Sgt. Thomas Lawson.)

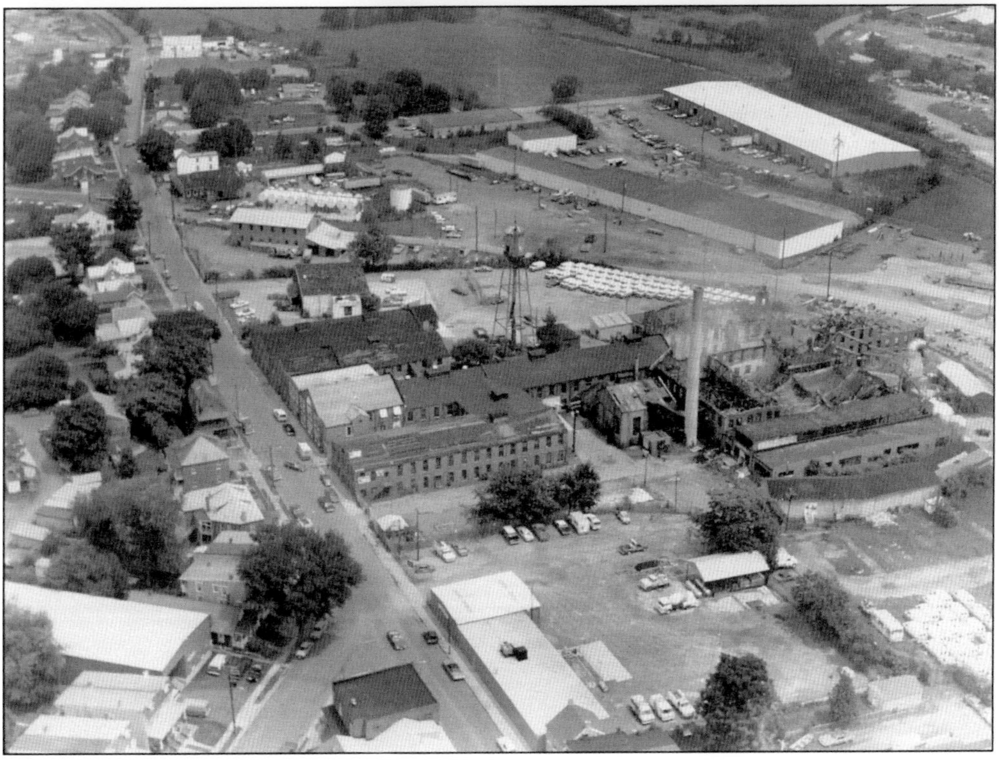

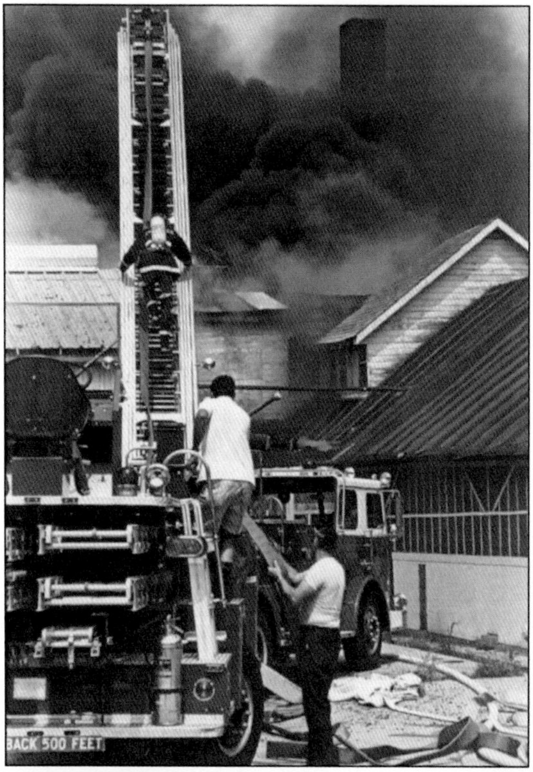

Fire races through the Jenkins Cannery buildings on South Street in August of 1979. (Courtesy Sgt. Thomas Lawson.)

An explosion and fire at the Wash Tub on South Bentz Street threatened several structures. From left to right, Robert Young, Terry Shankle, and James Graham of the Uniteds race to get hose lines in place. (Courtesy UFC.)

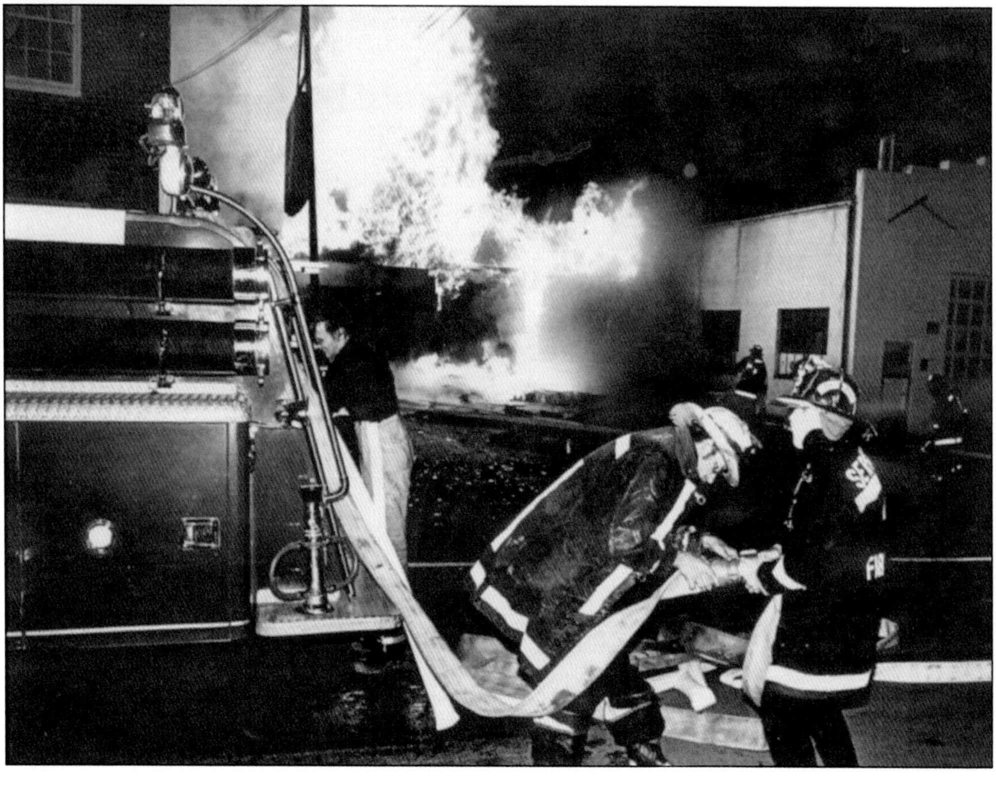

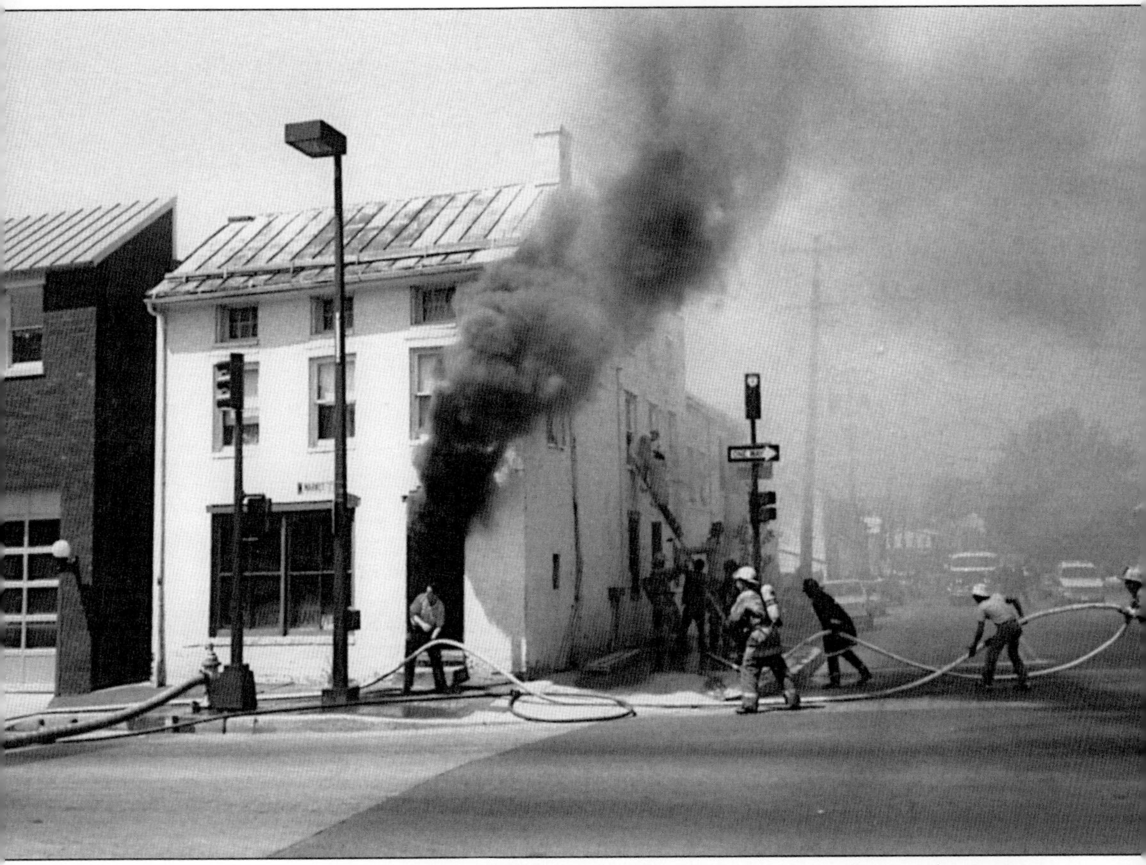

On a Sunday morning in the summer of 1988, Sgt. John Hahn was on duty at the Juniors when he heard someone yelling from the street in front of the station. The citizen said that the restaurant next to the station was on fire. This picture shows firefighters attacking the fire as a simultaneous rescue took place on the 6th Street side of the fire building. Fortunately, there was minimal damage to the north bay of the Juniors. (Collection of Charles M. Hahn.)

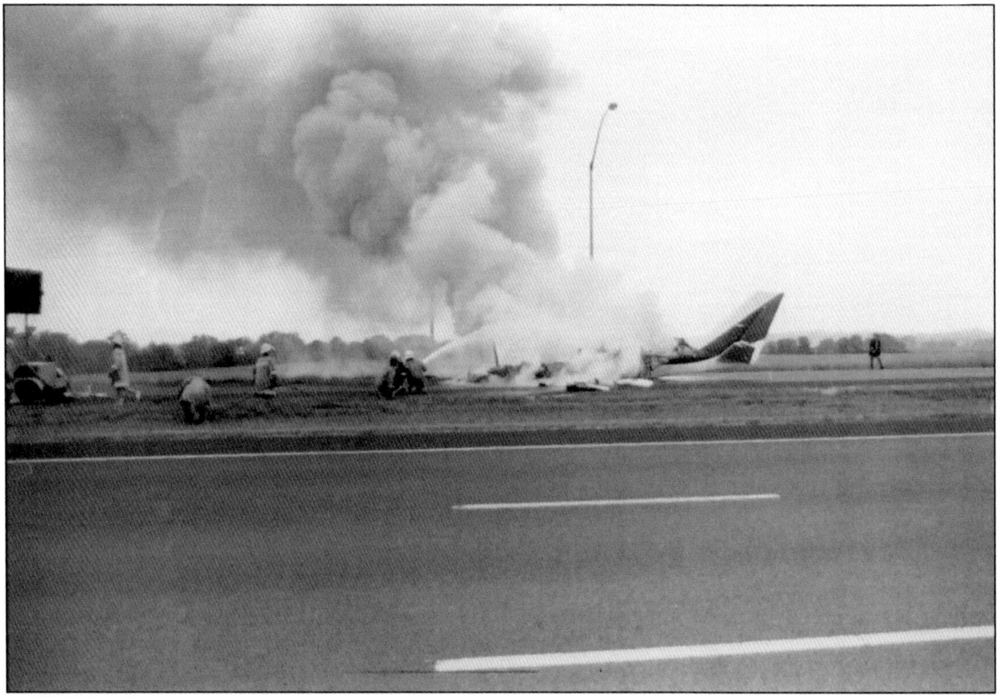

Fortunately, no one on the highway was injured when a plane crash landed on I-70 near the Frederick airport. (Collection of Charles M. Hahn.)

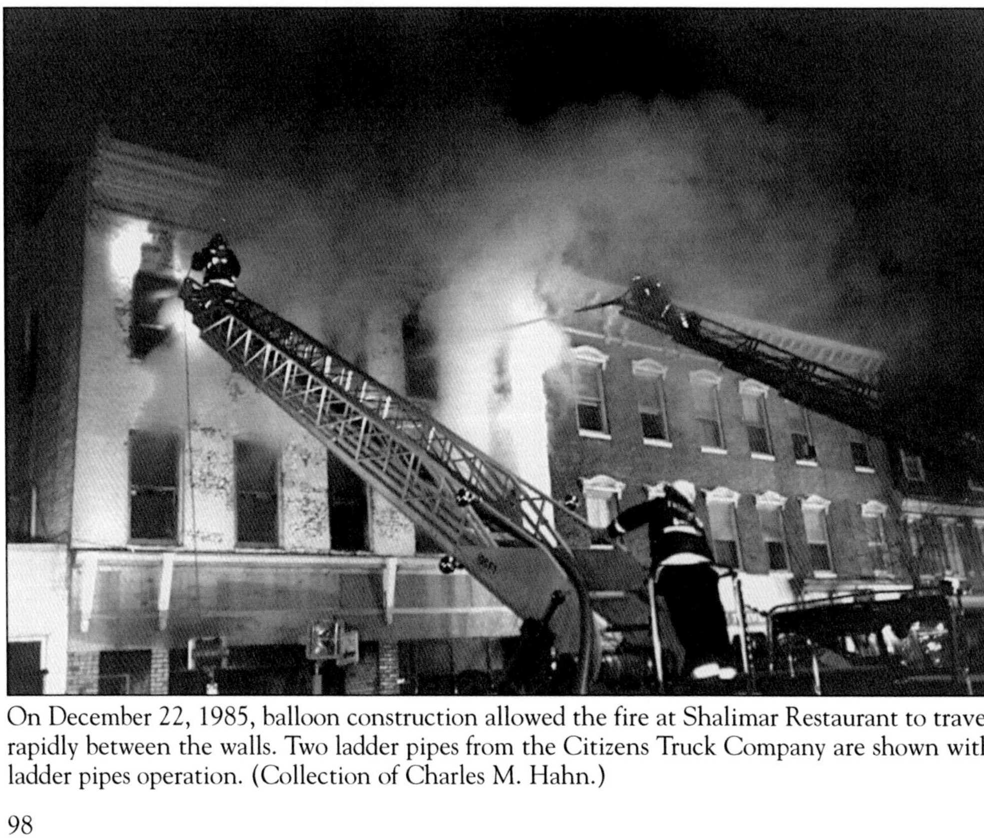

On December 22, 1985, balloon construction allowed the fire at Shalimar Restaurant to travel rapidly between the walls. Two ladder pipes from the Citizens Truck Company are shown with ladder pipes operation. (Collection of Charles M. Hahn.)

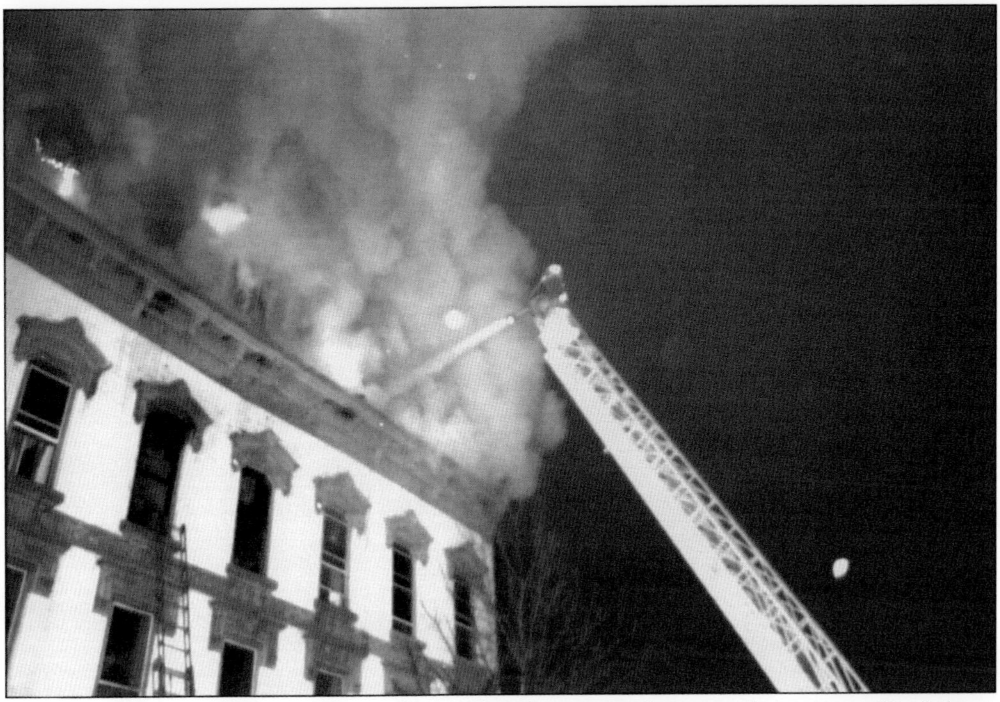

The Winebrenner Building was a former hotel converted to condominiums. Firefighters initially attempted an interior attack but eventually required the ladder pipe of the Citizens Truck Company to help control the blaze. (Collection of Charles M. Hahn.)

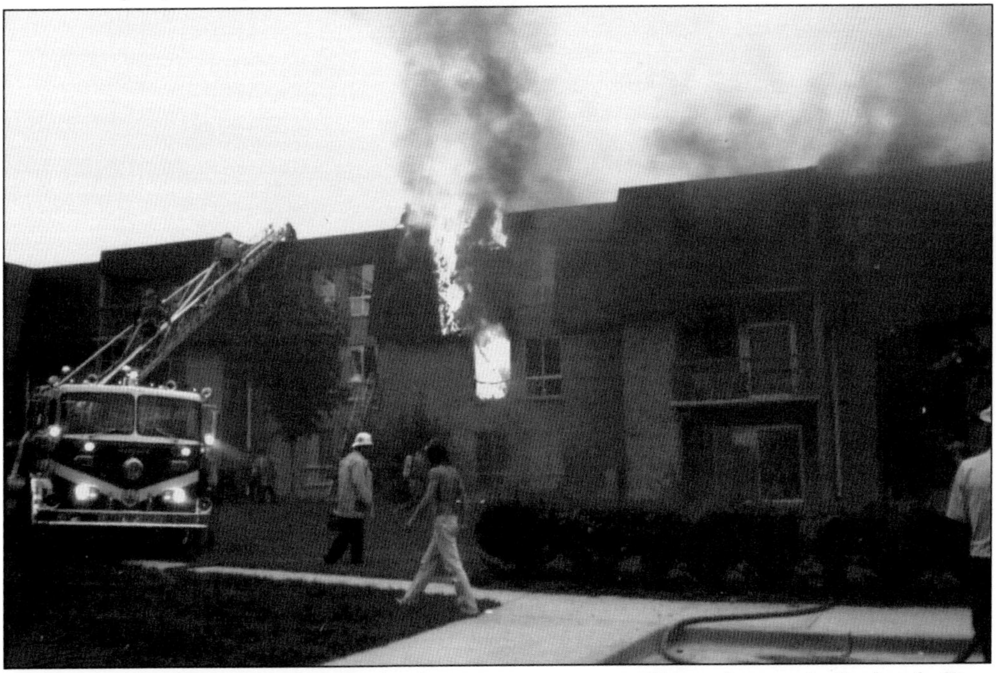

Fire breaks through the roof at the Applegate apartments on Taney Avenue in Frederick. Fires in garden apartments are becoming more prevalent as Frederick continues to grow. (Collection of Charles M. Hahn.)

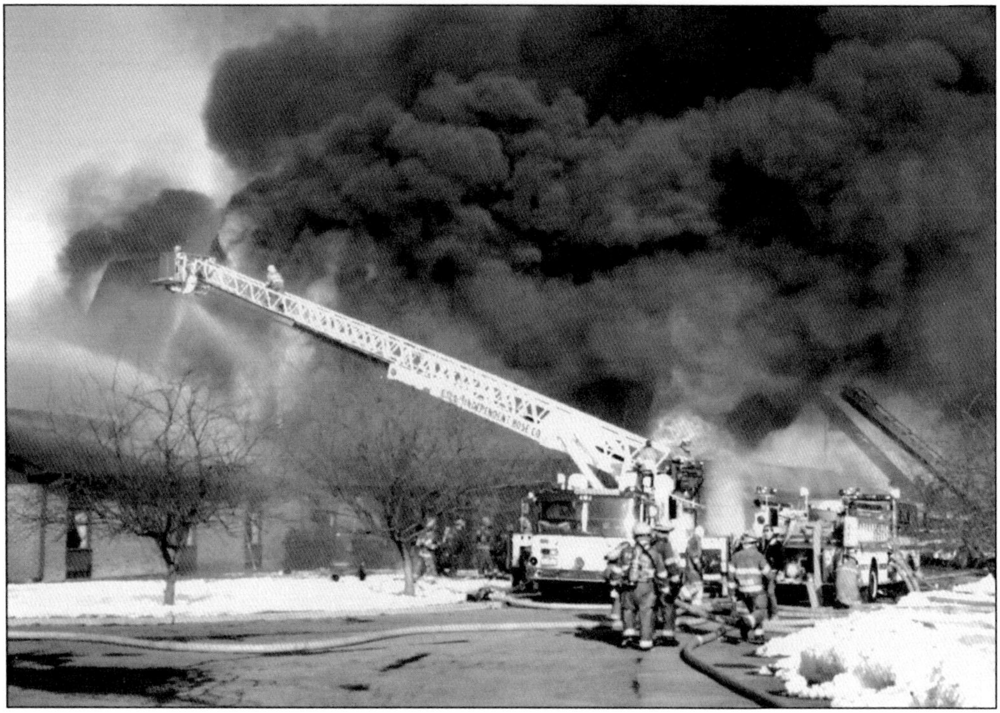

The Seventh Day Adventist Church fire occurred on February 28, 1993. Tower 1 begins ladder pipe operations as heavy fire blows through the roof. (Courtesy Warren Jenkins.)

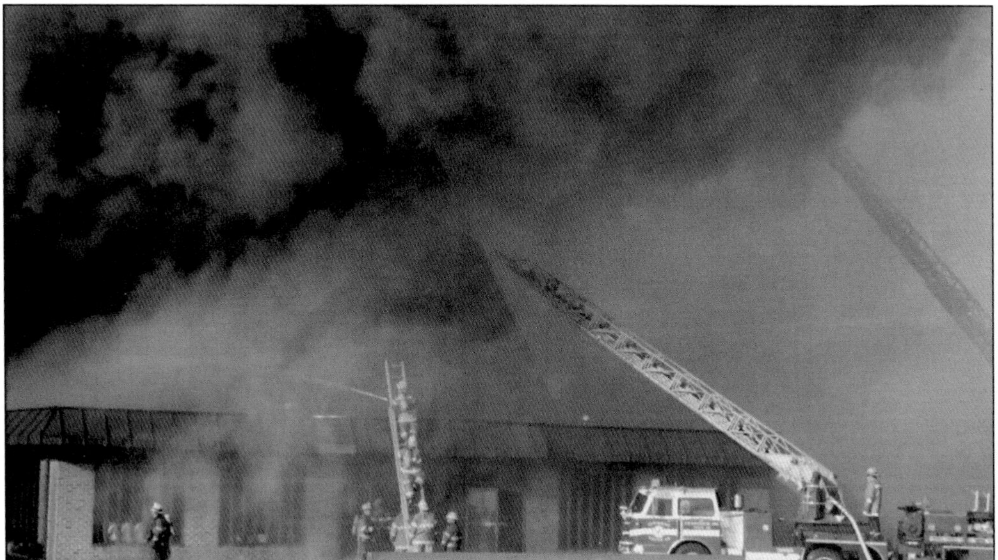

The ladder pipe of Truck 42 flows thousands of gallons of water at the height of the blaze. This truck received damage from the radiant heat of the fire. (Courtesy Warren Jenkins.)

Six

THE MARYLAND STATE FIREMEN'S ASSOCIATION IN FREDERICK

The first annual convention of the Maryland State Firemen's Association (MSFA) was called to order by Pres. J. Roger McSherry on June 7, 1893, at the City Opera House in Frederick. President McSherry, a member of the Independent Hose Company No. 1, called the meeting to order and in his opening remarks stated that the meeting "announces to the world, that the Volunteer Firemen of our State have at last awakened to the necessity of a more thorough and complete organization, and understanding this necessity, have realized that to promote and increase their efficiency as firemen, to encourage a fraternal feeling amongst one another and to secure sufficient protection for them as a class, such an organization is essential."

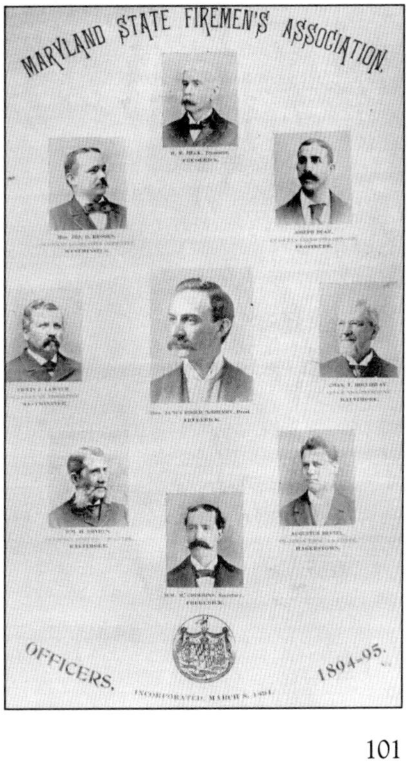

Shown here are the second-year officers of the Maryland State Firemen's Association in 1894. (Courtesy UFC.)

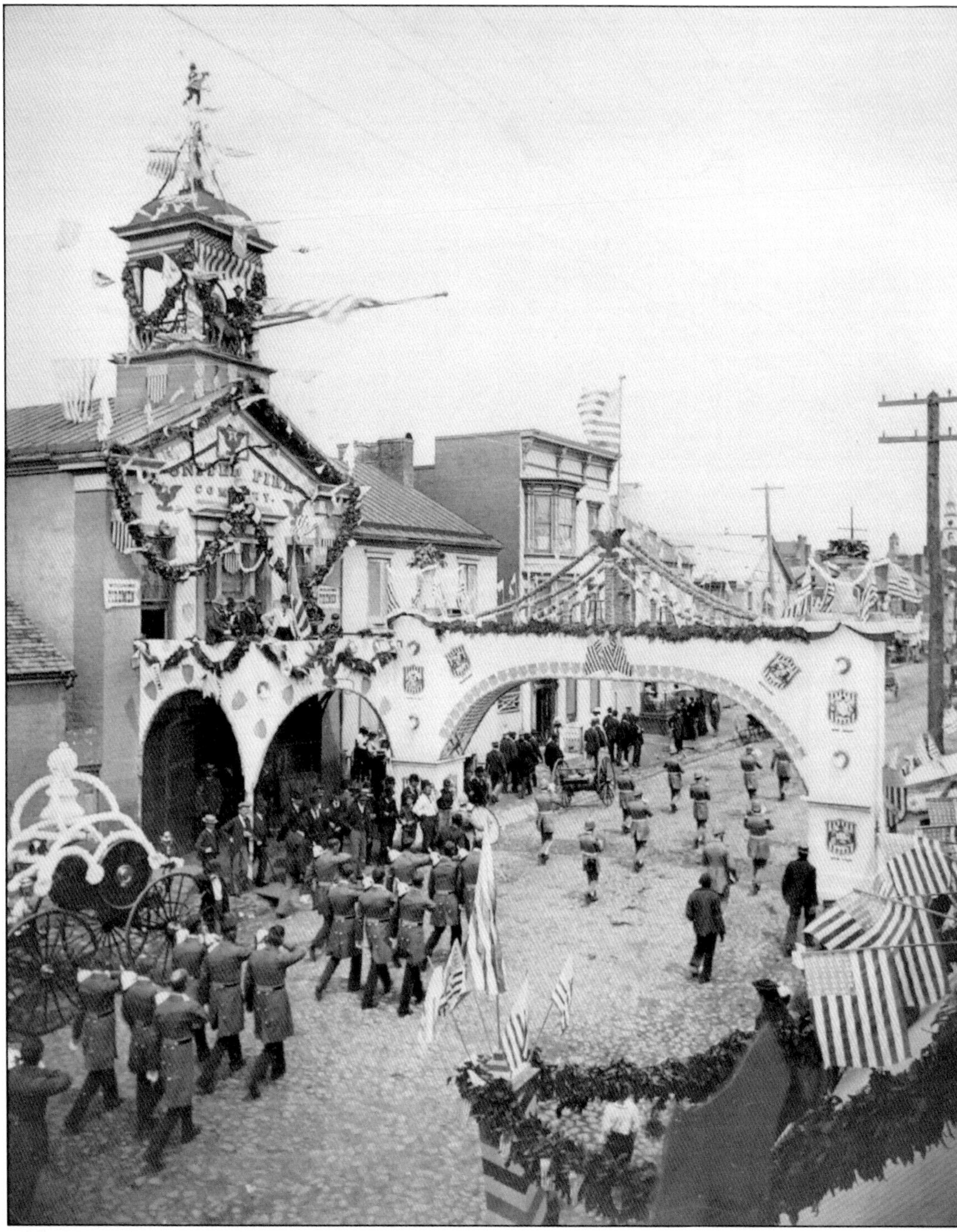

The United Fire Hall was decorated for the first convention of the MSFA in 1893. (Courtesy UFC.)

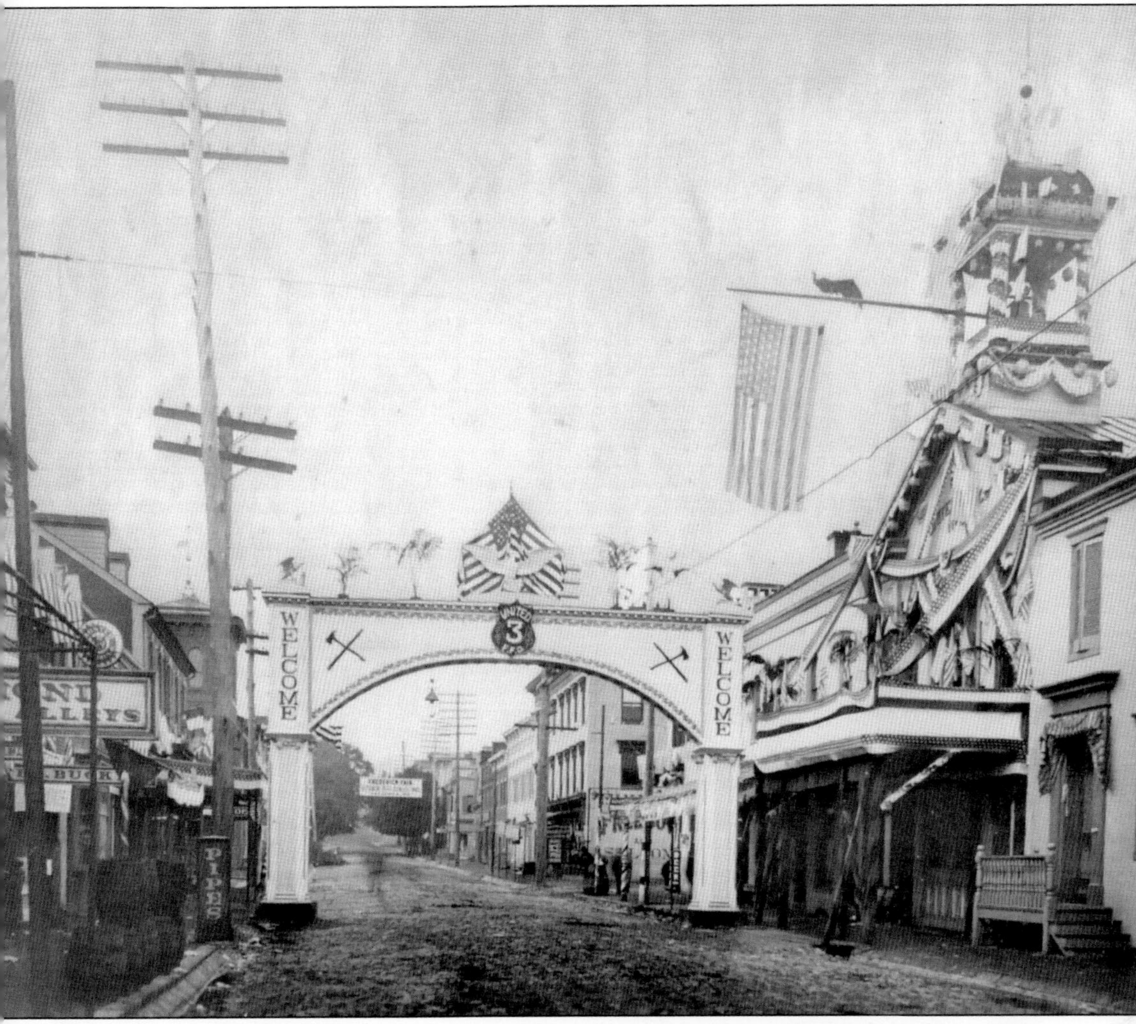

The Uniteds also provided a large arch for the next convention in Frederick, held in 1903. Note the turkey resting on the flag pole. (Courtesy UFC.)

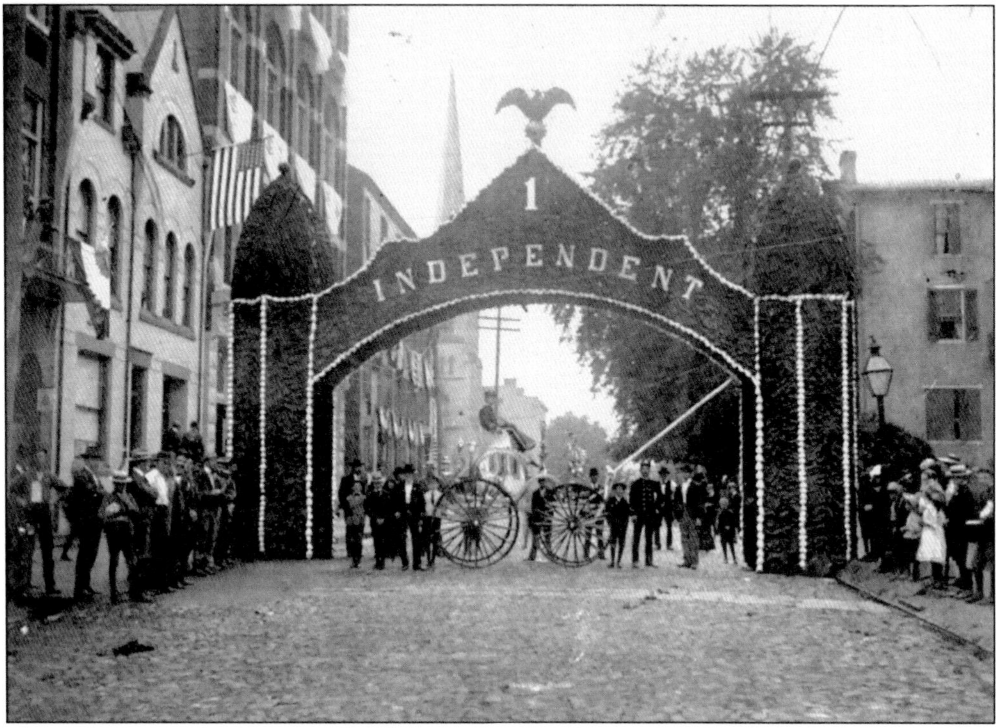

The Independents built a special arch in 1903. Note the 1893 Holloway Hose Carriage in the center of the arch. (Courtesy IHC.)

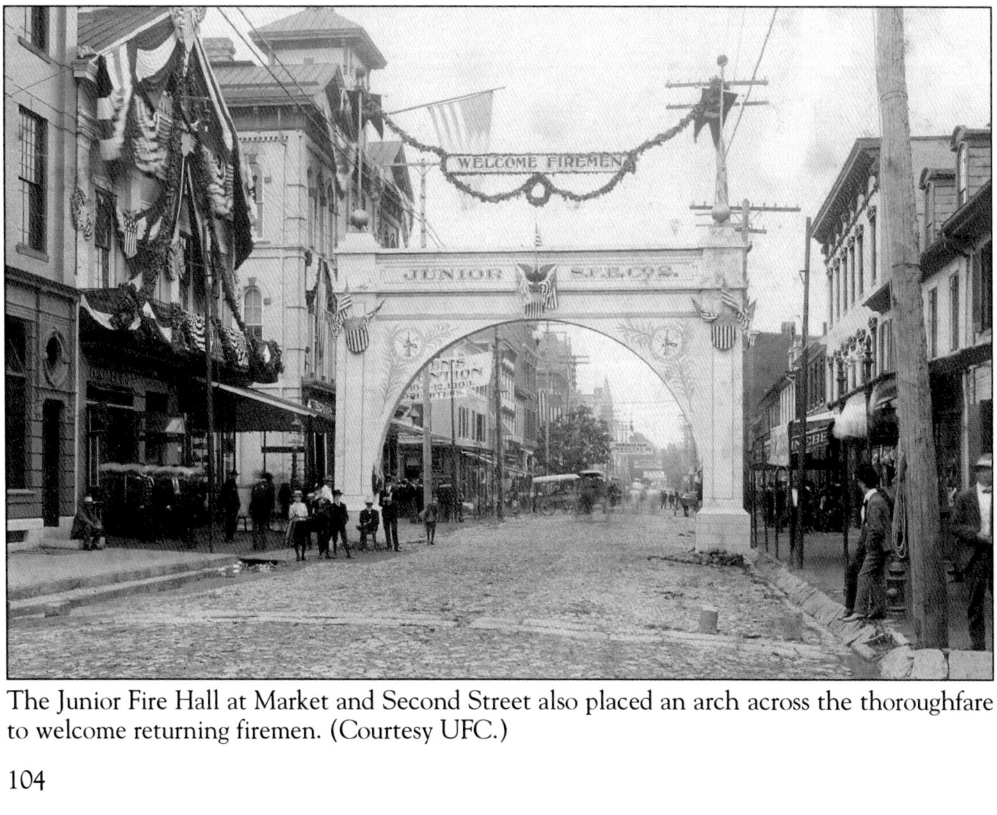

The Junior Fire Hall at Market and Second Street also placed an arch across the thoroughfare to welcome returning firemen. (Courtesy UFC.)

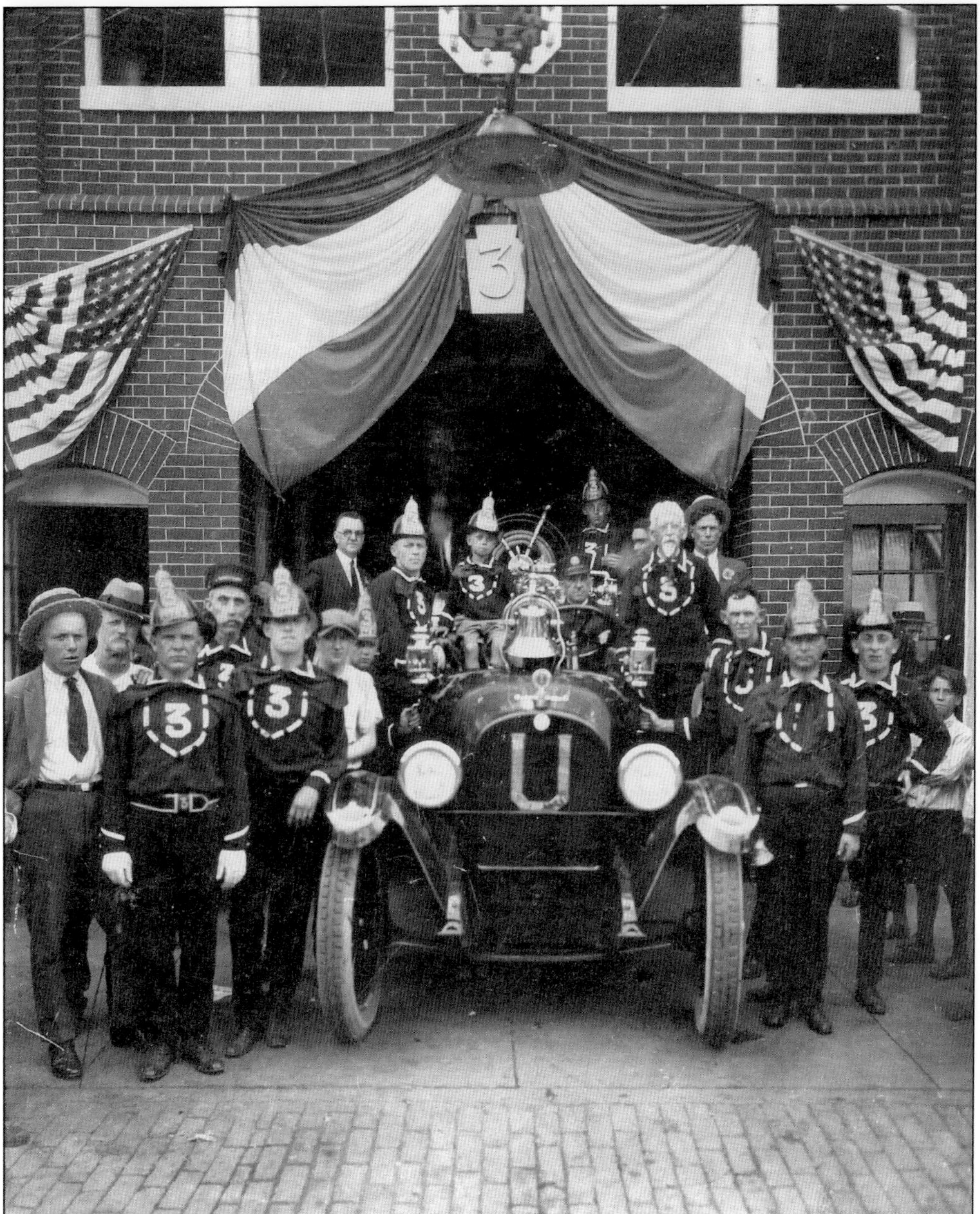

Members of the Uniteds pose in front of the enginehouse with the 1923 chemical/floodlight truck for the 1923 MSFA Convention. (Courtesy UFC.)

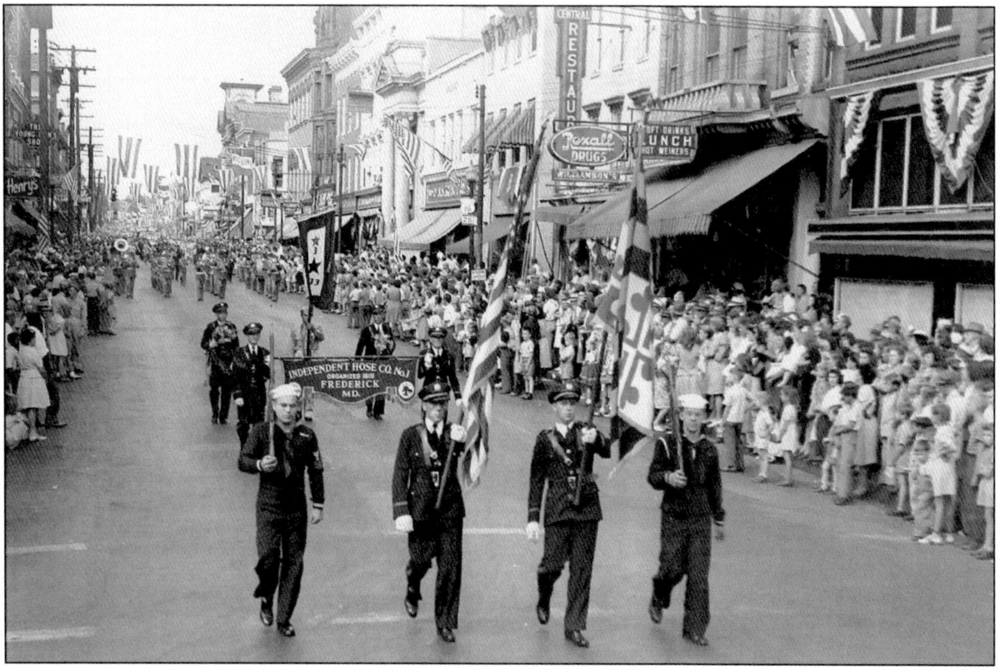

The first MSFA convention after World War II was held in Frederick in 1946. Hundreds of local residents gather along Market Street as the members of the armed forces joined firefighters from the Independent Hose Company to present the first colors of the parade. (Courtesy IHC.)

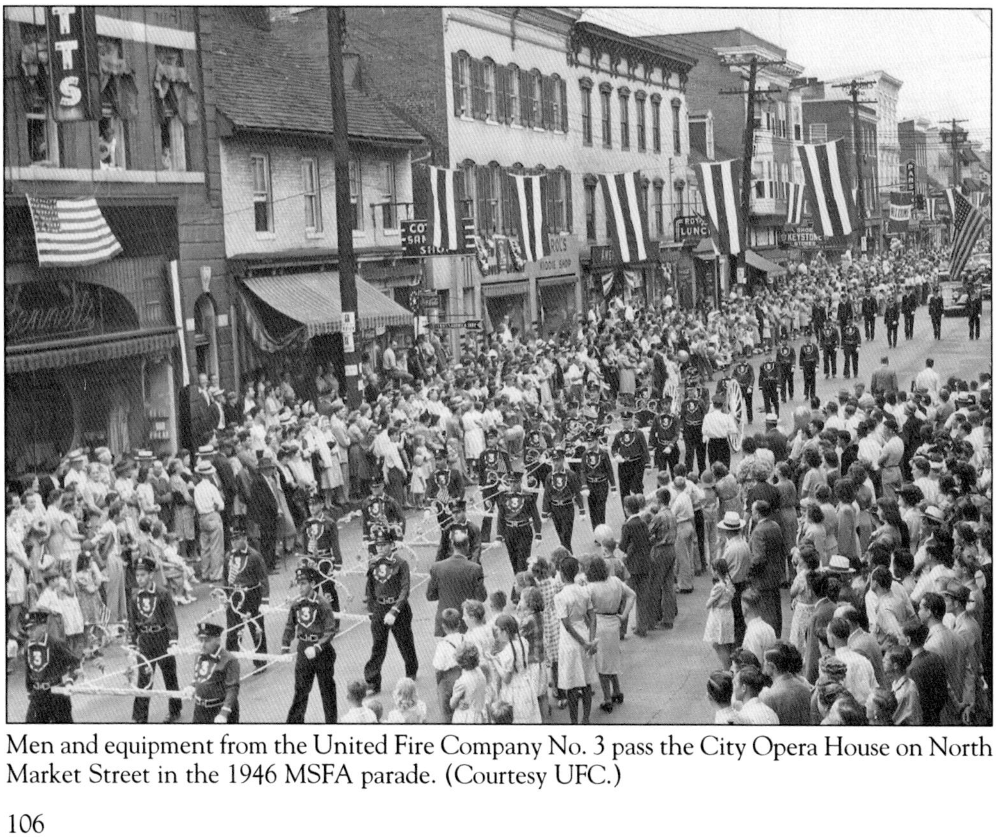
Men and equipment from the United Fire Company No. 3 pass the City Opera House on North Market Street in the 1946 MSFA parade. (Courtesy UFC.)

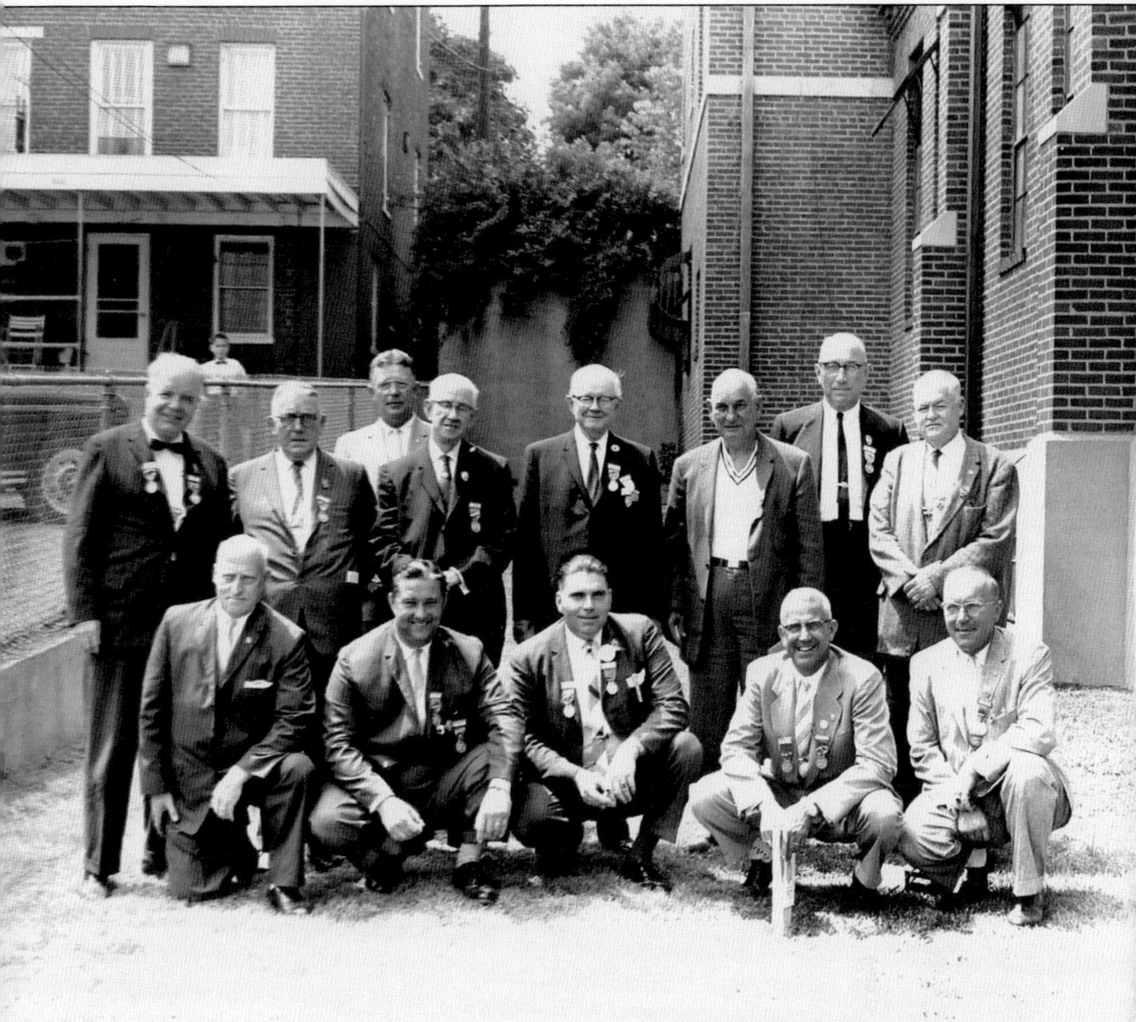

The past presidents of the Maryland State Firemen's Association, shown in this June 1962 photo, are, from left to right, (front row) Spencer J.H. Brown Sr. (1949–1950), Sandy Spring; John W. Smith (1954–1955), West Annapolis; Ralph N. Small Jr. (1959–1960), Boulevard Heights; Howard B. Springer (1958–1959), Havre DeGrace; and Cromwell C. Zembower (1951–1952), LaVale; (back row) Ted C. Gardner (1955–1956), Riverdale Heights; B. Harrison Shipley (1937–1938), Ellicott City; W. Bartgis Storm (1946–1947), Frederick; Gov. J. Millard Tawes (1936–1937), Crisfield; Vincent A. Simmel (1929–1930), Cottage City; G. Mitchell Boulden (1950–1951), Elkton; Philip H. Beard (1956–1957), Walkersville; and William A. Chenoweth (1953–1954), Pikesville. (Courtesy UFC.)

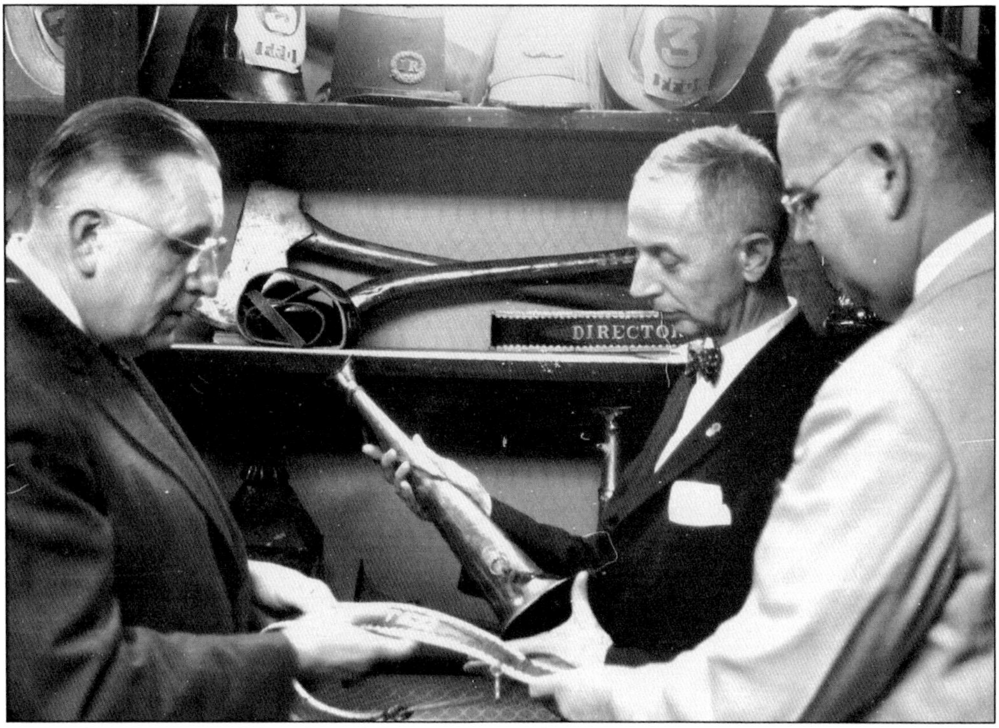

Pres. Alton Bennett of the United Fire Company shows MSFA past presidents Howard Keller (1961–1962) of Pikesville and Jack V. Reckner (1960–1961) of Earleigh Heights some of the historic artifacts in the board room of the Uniteds. (Courtesy UFC.)

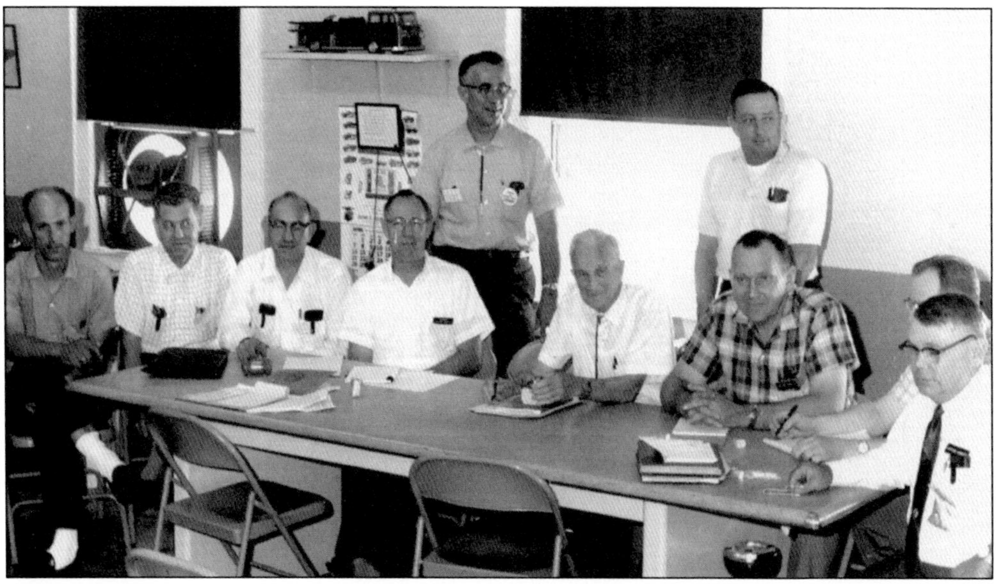

MSFA president William Moore (1967–1968), United Fire Company No. 3 of Frederick, is at an executive meeting held at the Independent Hose Company. Shown from left to right are (seated) Charles Shindle, Francis Fatkin, George Mayer, Melvin Schwearing, Floyd Heimer, Daniel Smith, Charles Burton, and Buck Gladhill; (standing) President Moore and Thomas L. Reynolds. (Courtesy UFC.)

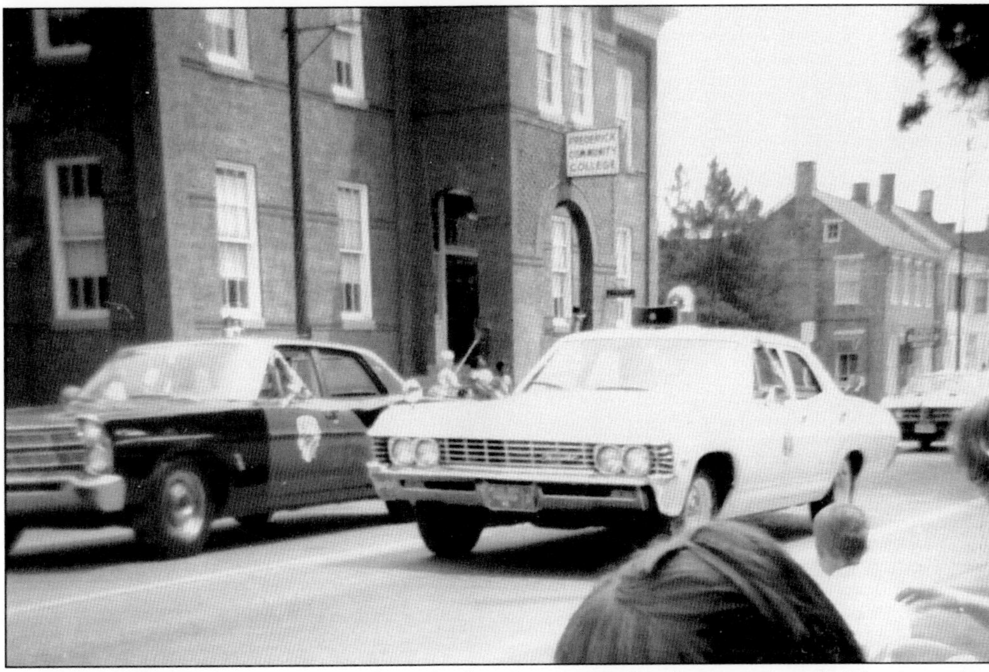

The 1967 MSFA Convention parade in Frederick was over four hours in length. Vintage police cruisers from the Frederick Police Department and Maryland State Police lead the parade through downtown Frederick. (Courtesy H. Richard Hahn.)

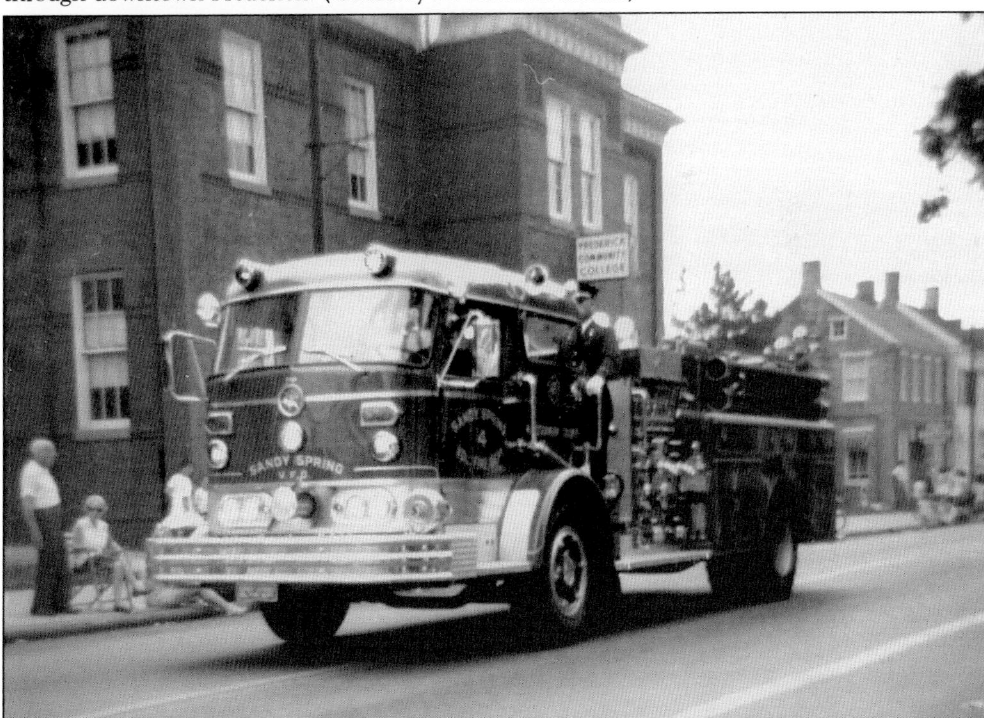

A Mack pumper from the Sandy Spring Volunteer Fire Department in Montgomery County drives in the 1967 convention parade. (Courtesy H. Richard Hahn.)

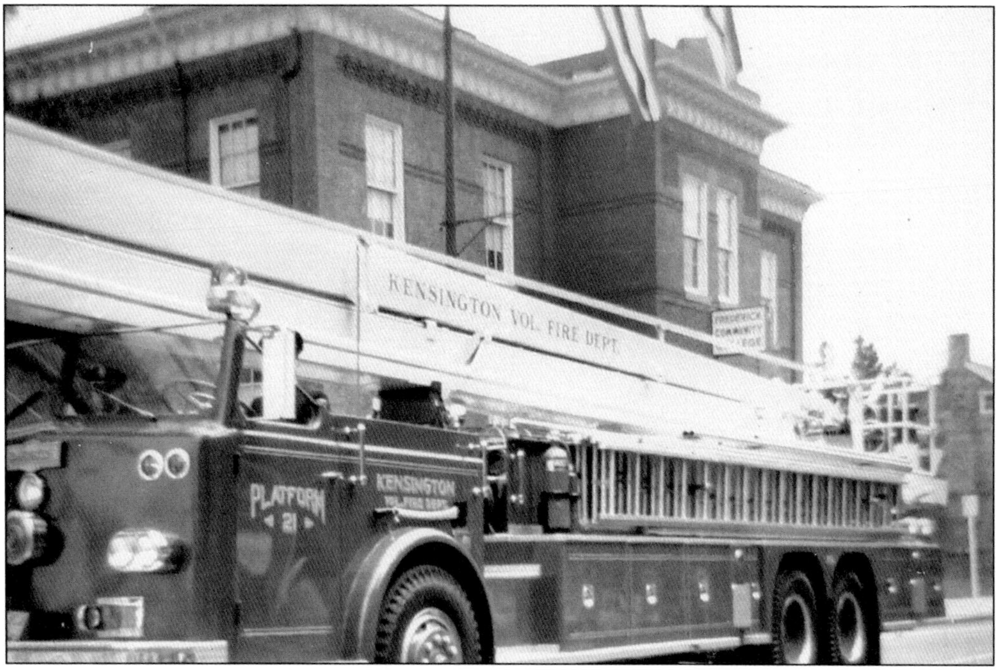

The Kensington Volunteer Fire Department proudly displays Platform 21, one of the first tower/aerial devices in Maryland, at the 1967 convention parade in Frederick. (Courtesy H. Richard Hahn.)

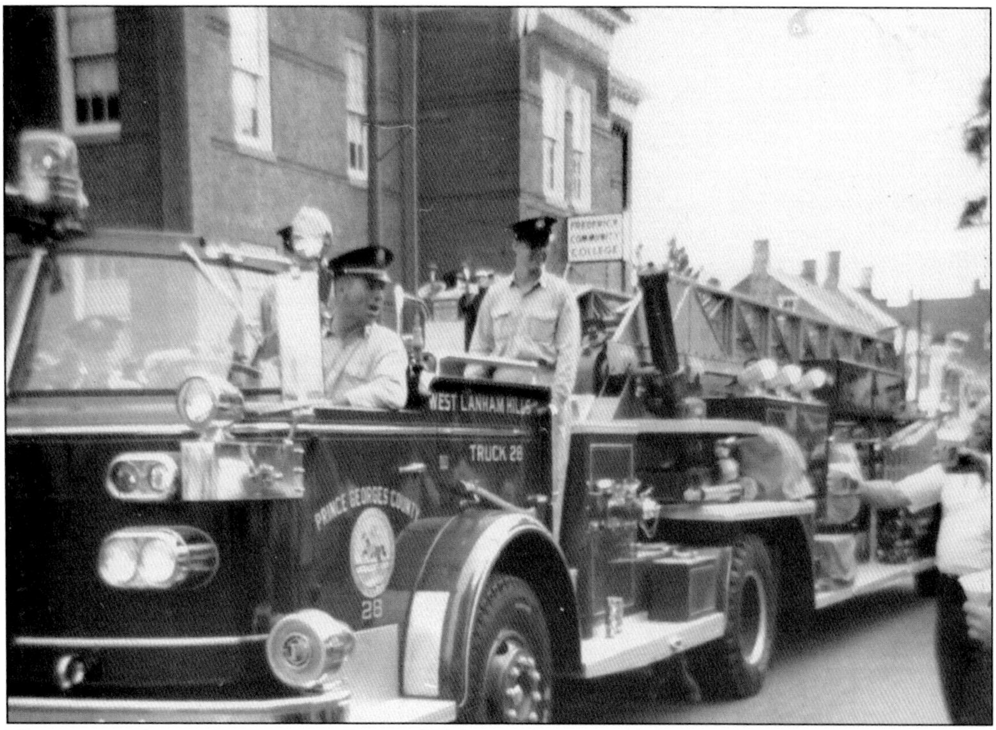

This tractor-drawn American LaFrance ladder truck came all the way from West Lanham Hills in Prince George's County to participate in the 1967 parade. (Courtesy H. Richard Hahn.)

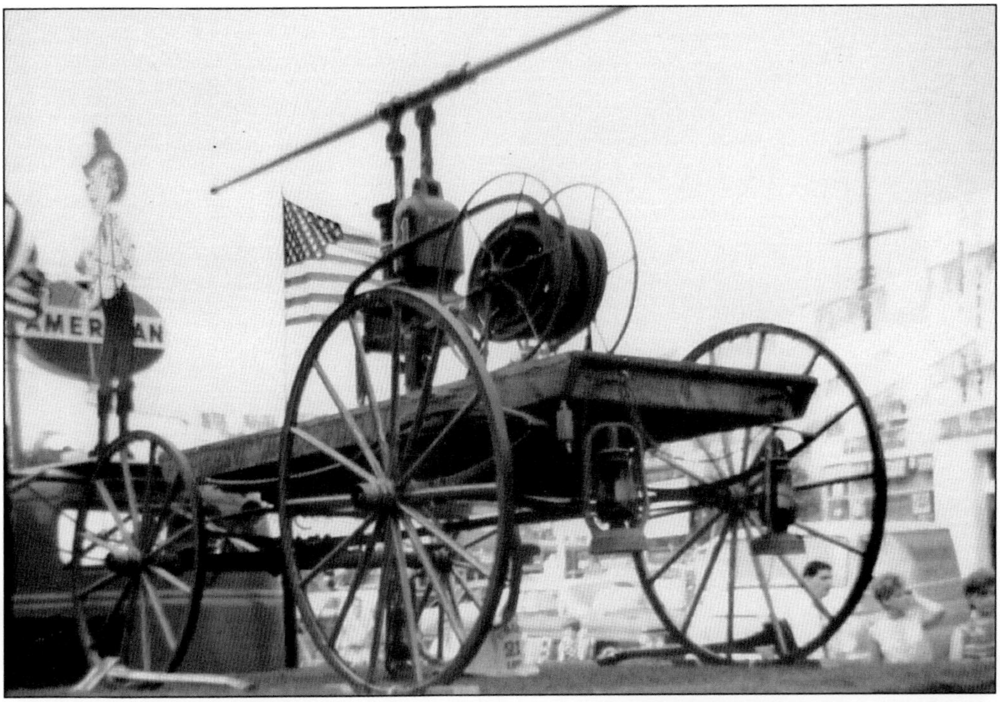

An antique hand-pumper is displayed during the 1967 convention parade in Frederick. (Courtesy H. Richard Hahn.)

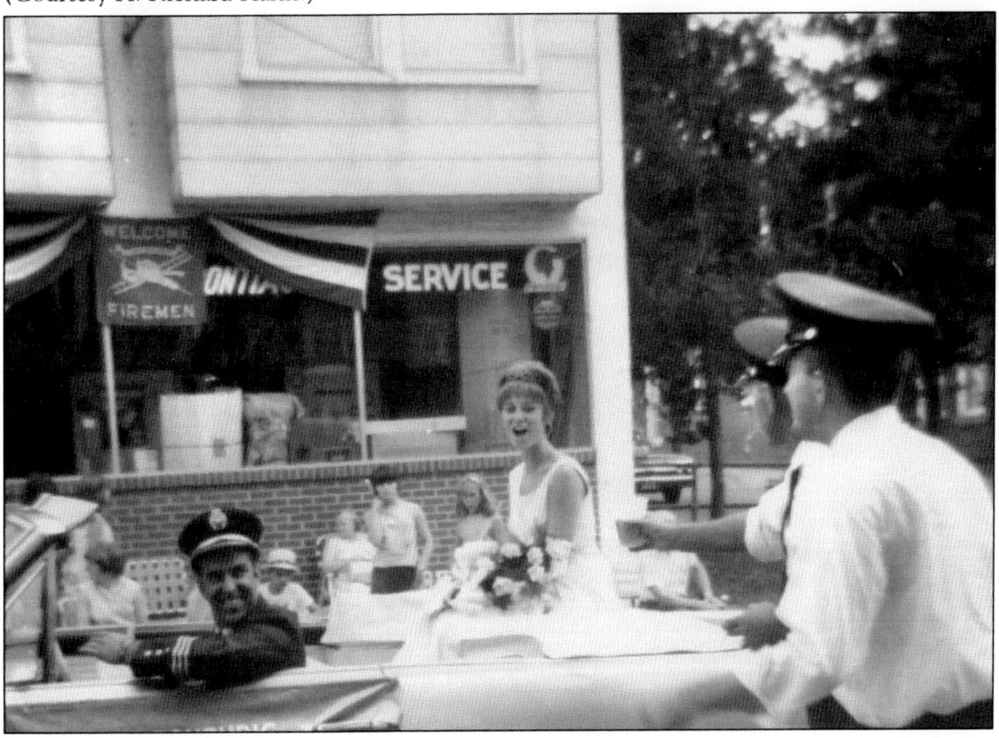

The Fire Prevention Queen from the Potomac Heights Volunteer Fire Department of Charles County rides in the 1967 convention parade. (Courtesy H. Richard Hahn.)

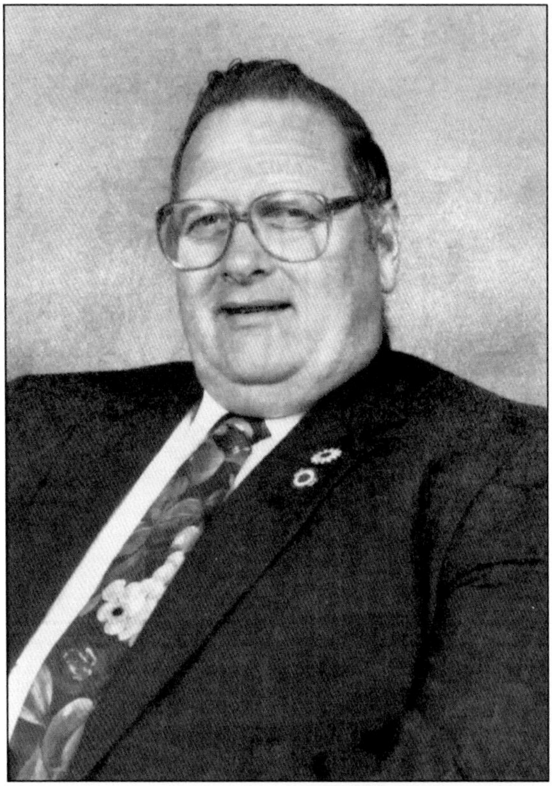

It was almost 25 years later before another MSFA president was from Frederick. Mr. Richard L. Yinger of the Citizens Truck Company held the office in 1993–1994.

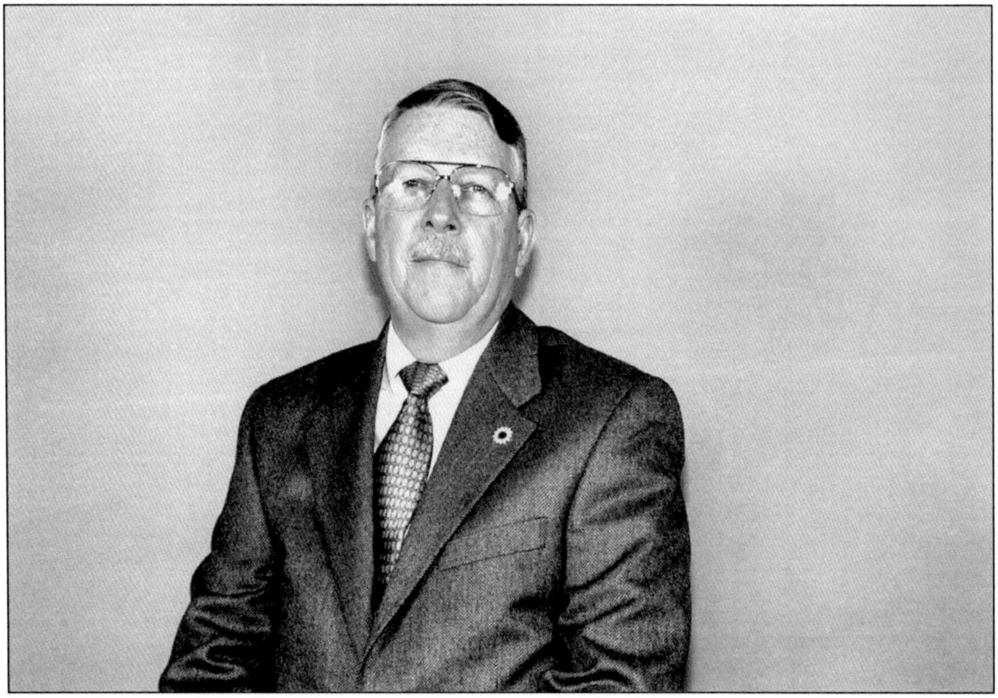

The Citizens Truck Company again provided the MSFA president in 2004 when Robert Jacobs was elected to lead the association.

Seven

ODDS AND ENDS

The final chapter highlights areas that apply to several companies or, in some cases, companies or apparatus from outside the Frederick area—some very far outside!

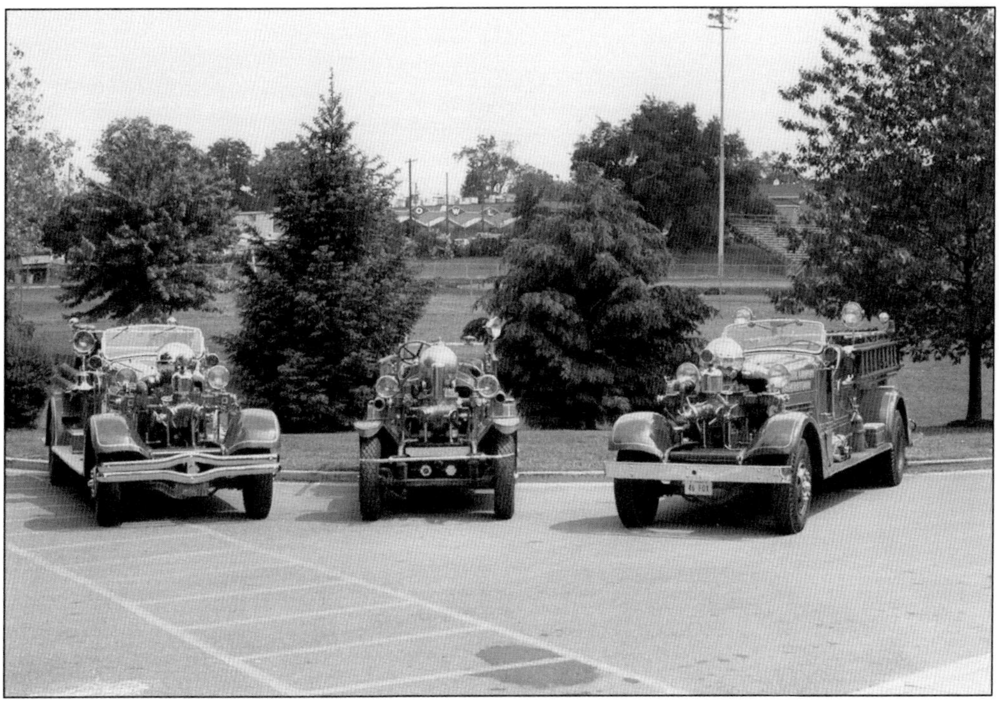

The Foxes from Frederick and Hagerstown meet many years after retirement at a Frederick muster. The United 1939 Ahrens-Fox is on the left, the 1919 Ahrens-Fox formerly owned by the Uniteds is in the middle, and the 1946 Ahrens-Fox of First Hose of Hagerstown is on the right. (Courtesy Warren Jenkins.)

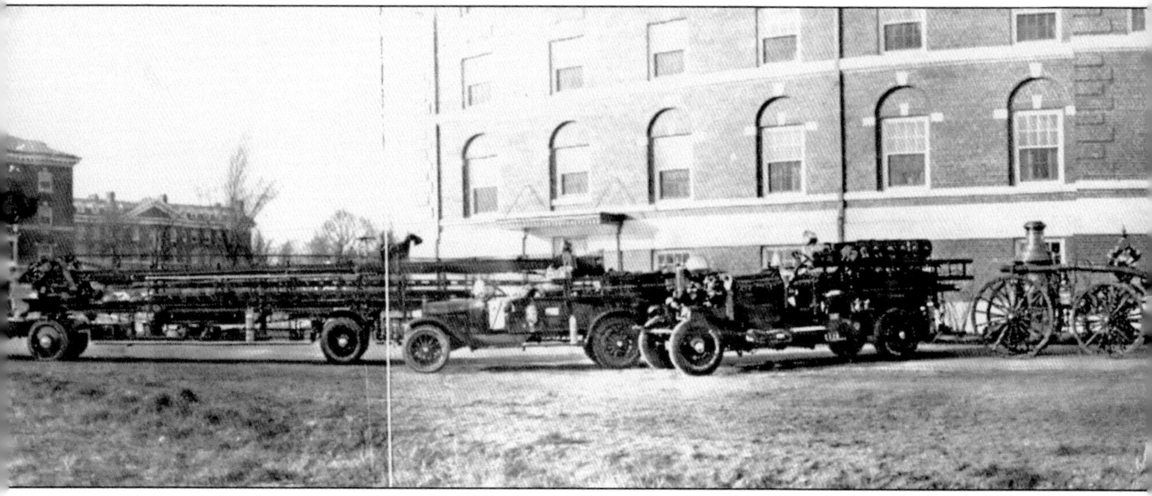

All equipment in the city of Frederick is proudly displayed in front of the dormitory at Hood College. From left to right are Citizen Truck Company 1927 Ahrens-Fox ladder truck, United Fire Company No. 3 1923 chemical/combination engine, United 1919 Ahrens-Fox, United

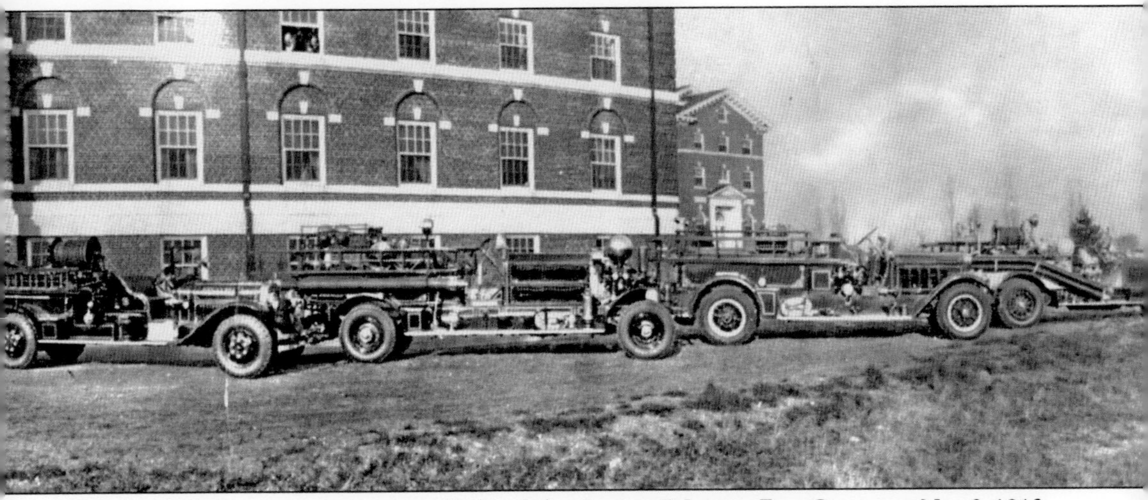

1878 Clapp & Jones steam pumper "Lily of the Swamp," Junior Fire Company No. 2 1913 White LaFrance chemical engine, Junior 1924 Ahrens-Fox, Independent Hose Company No. 1 1931 American LaFrance, and Independent 1921 American LaFrance. (Courtesy UFC.)

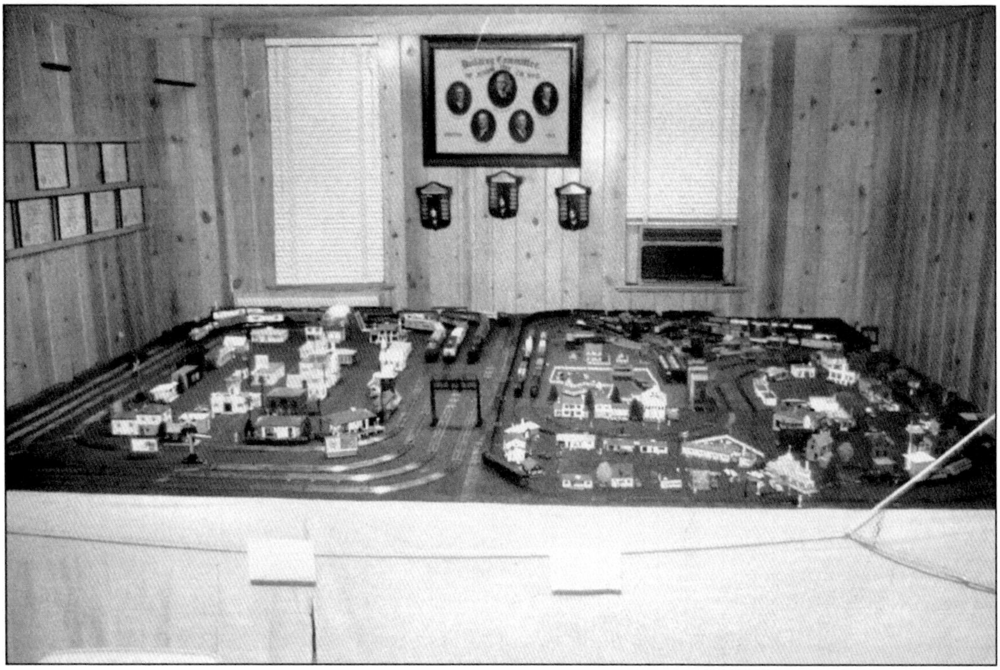

Train gardens have been a firehouse favorite. For many years, the Citizens Truck and the Junior Fire Companies have displayed trains around the Christmas holidays. Here is a display in the engine house of the Juniors that used trains from several members. (Courtesy H. Richard Hahn.)

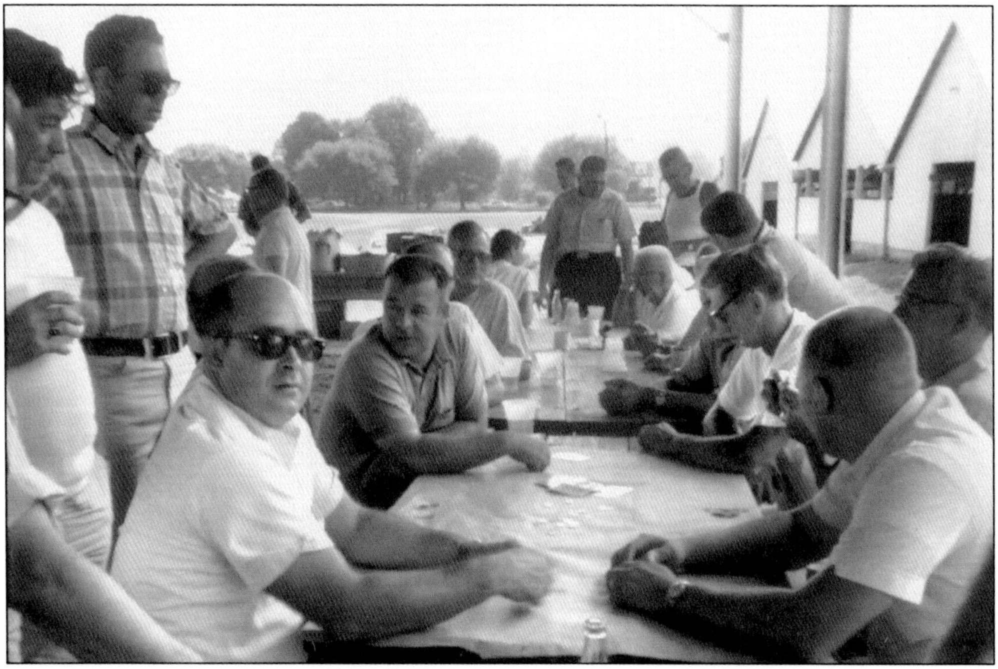

A picnic was held for many years to bring the members of the city companies together. This picture shows Sgt. Charles "Mutt" Deater, longtime driver at the Uniteds, dealing one of his famous card games. (Courtesy H. Richard Hahn.)

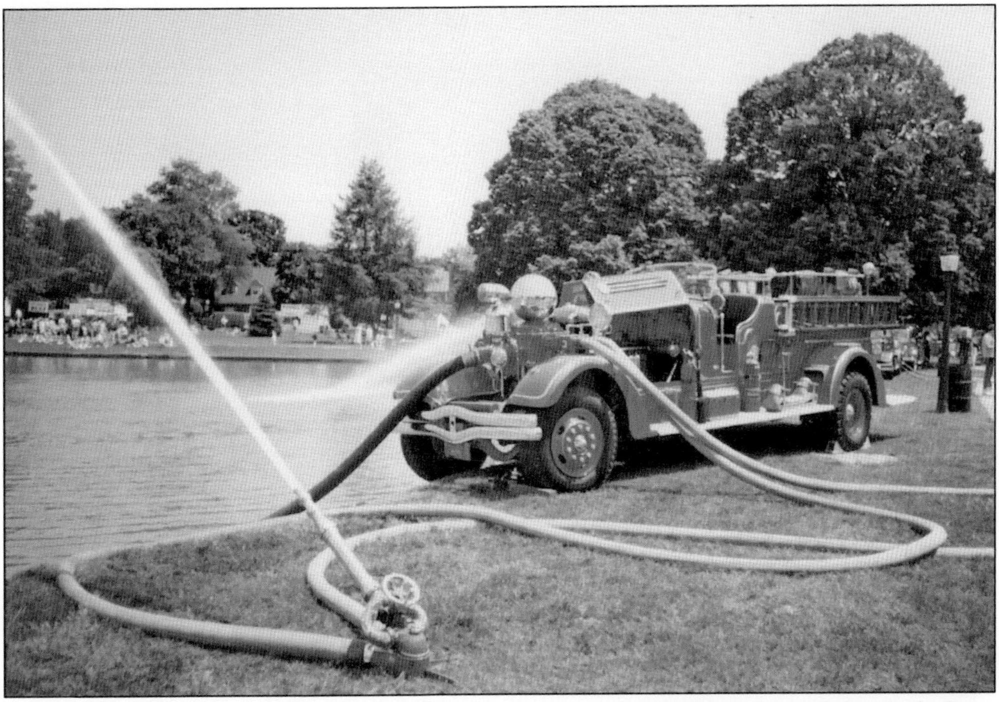

Antique fire apparatus musters were popular for many years at Culler Lake in Baker Park. Even though the engine was nearly 50 years old, the United 1939 Ahrens-Fox shows it can still pump with the best of the new during a pumping contest. (Courtesy Warren Jenkins.)

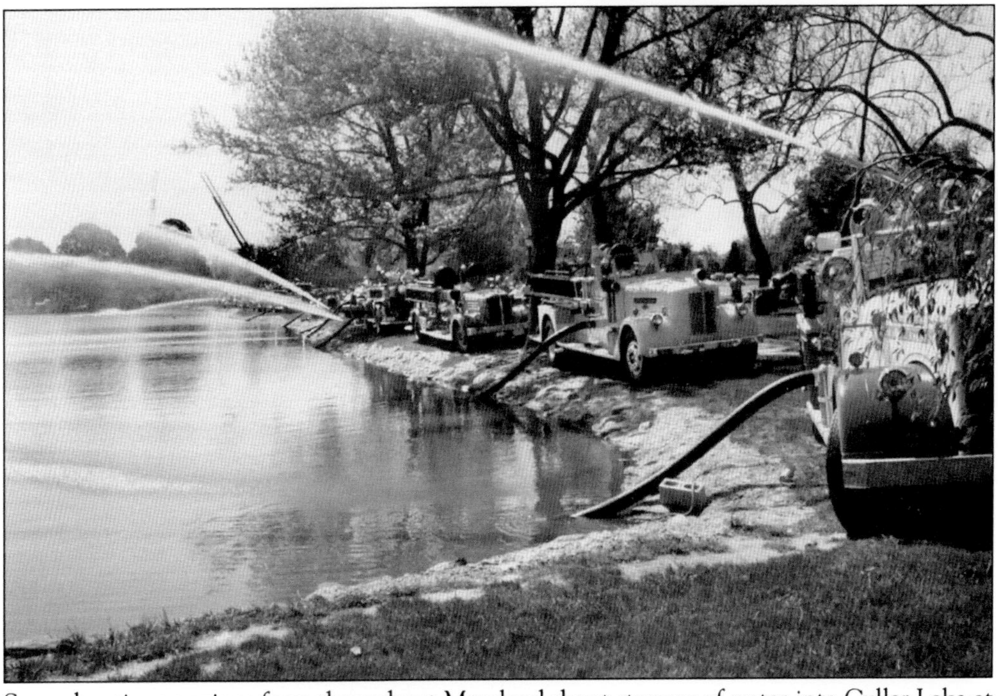

Several antique engines from throughout Maryland shoot streams of water into Culler Lake at an annual muster. (Courtesy Warren Jenkins.)

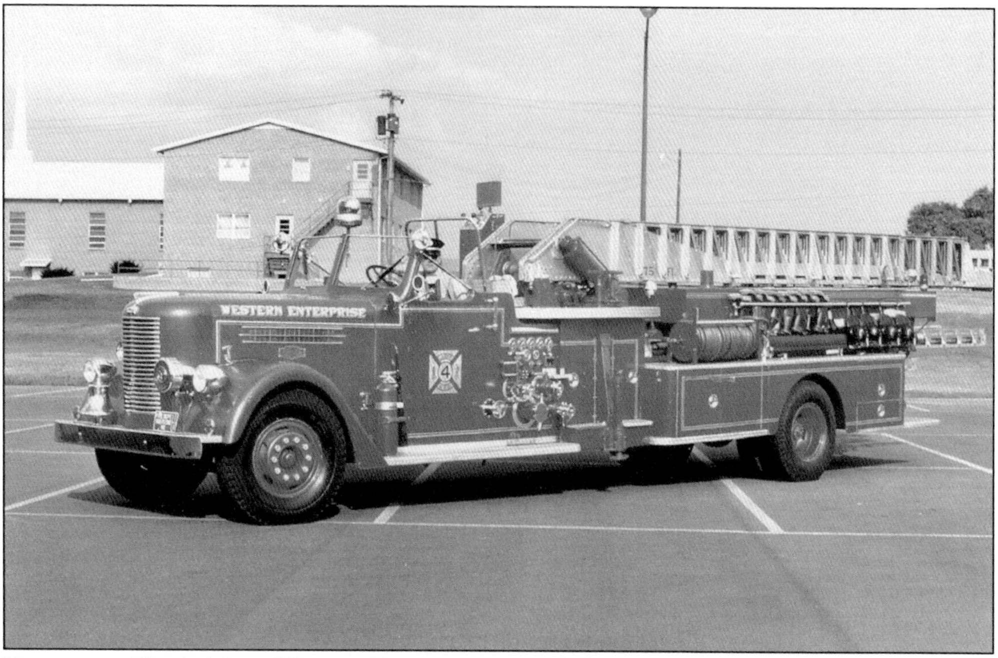

Hagerstown Truck 4 from Western Enterprise responded to many major fires in Frederick until a second ladder truck was purchased in 1974. Without hesitation, this truck responded on a 25-mile, one-way run from Hagerstown over two mountains to assist Frederick firefighters. (Courtesy Warren Jenkins.)

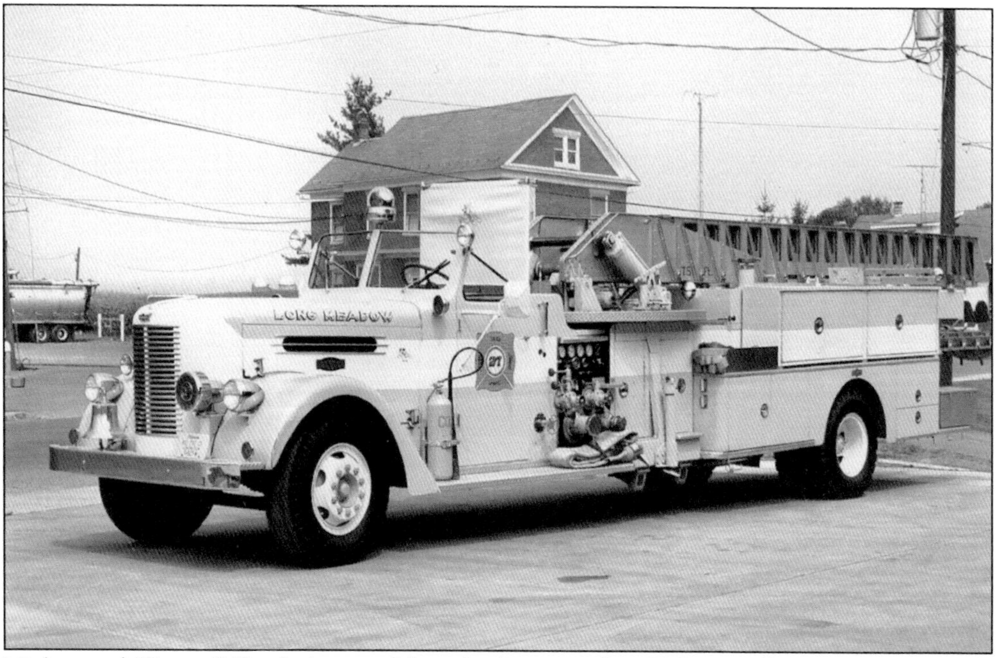

Nicknamed "Bruno," Truck 4 was sold to the Longmeadow Volunteer Fire company outside of Hagerstown and eventually returned to Frederick when George Fulmer, a member of the Independents and an avid collector, purchased the truck and showed it at many local parades. (Courtesy Warren Jenkins.)

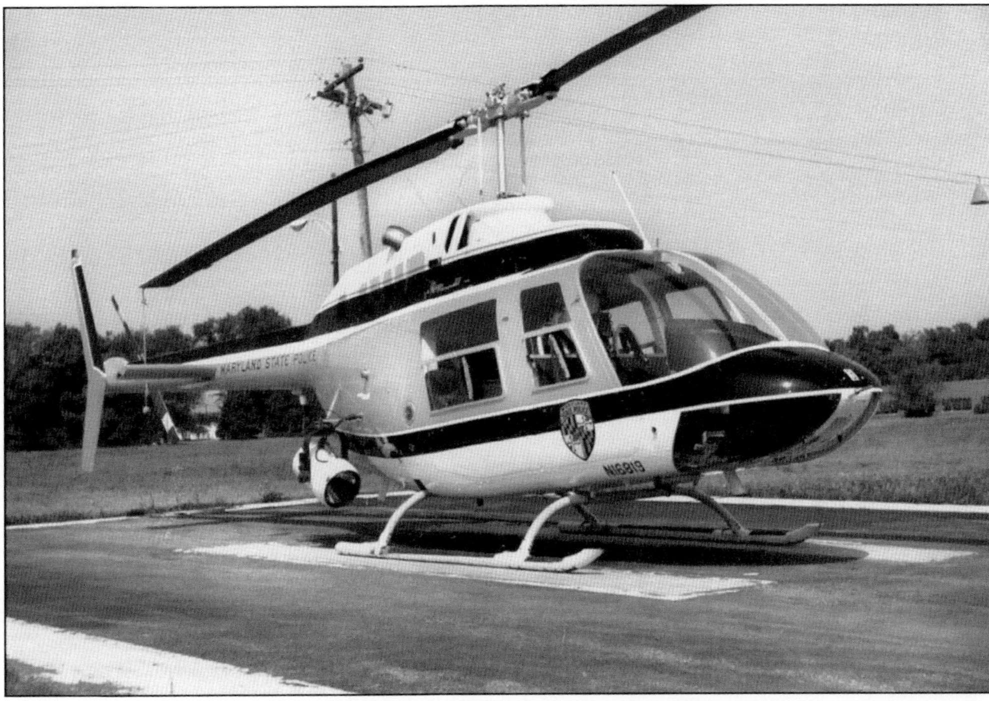

The Maryland State Police utilized a Bell Jet Ranger when they opened a hangar at the Frederick airport in 1973 to provide a medi-vac helicopter to serve the Frederick area. (Courtesy Warren Jenkins.)

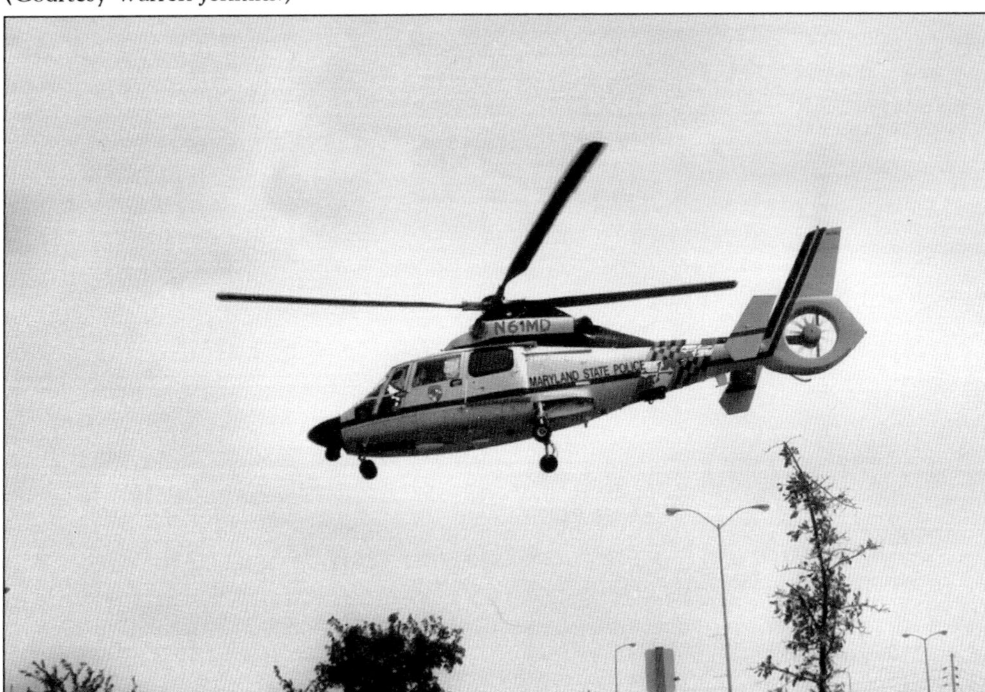

In 1989, the Maryland State Police helicopter fleet was modernized, and larger Dauphine helicopters replaced the small Bell Jet Rangers.

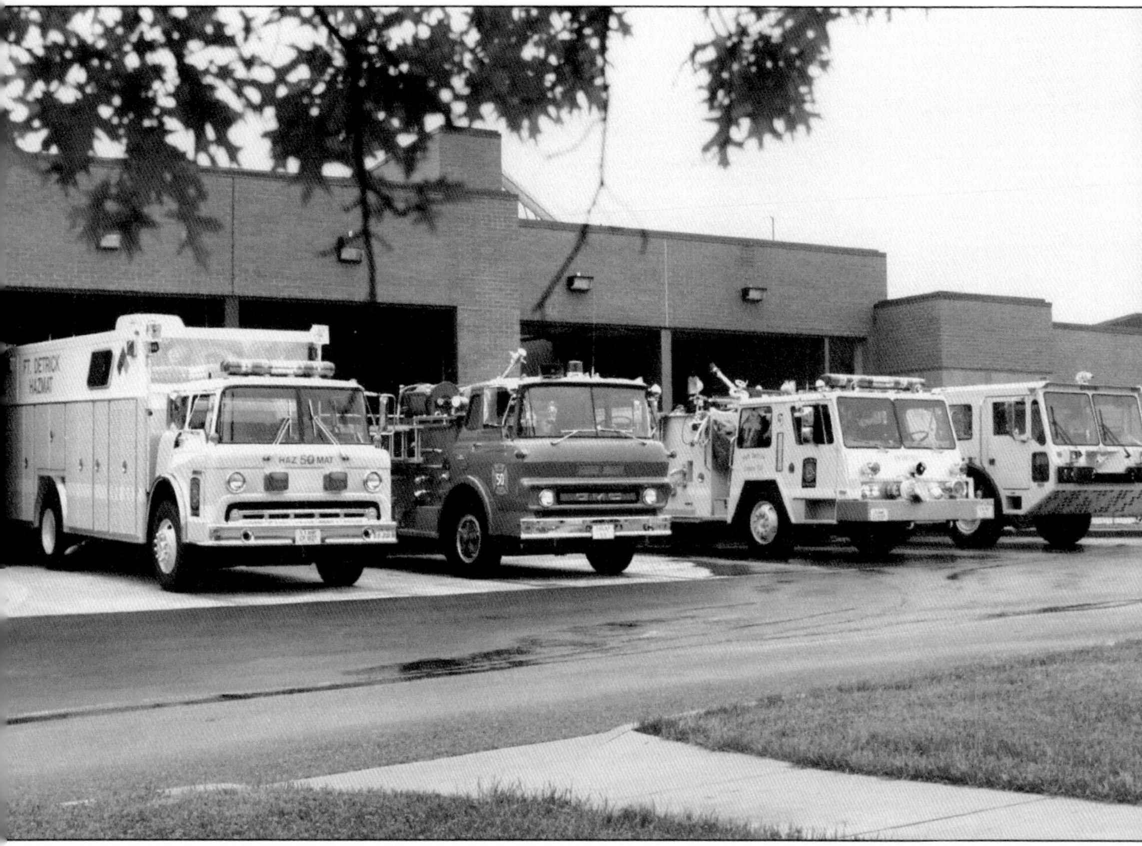

Since 1950, the City of Frederick has had a written mutual aid agreement with the Fort Detrick Fire Department (Company 50). For years, Company 50 only responded off post for major fires. Now, they respond automatically to alarms where they have a closer response. (Courtesy Warren Jenkins.)

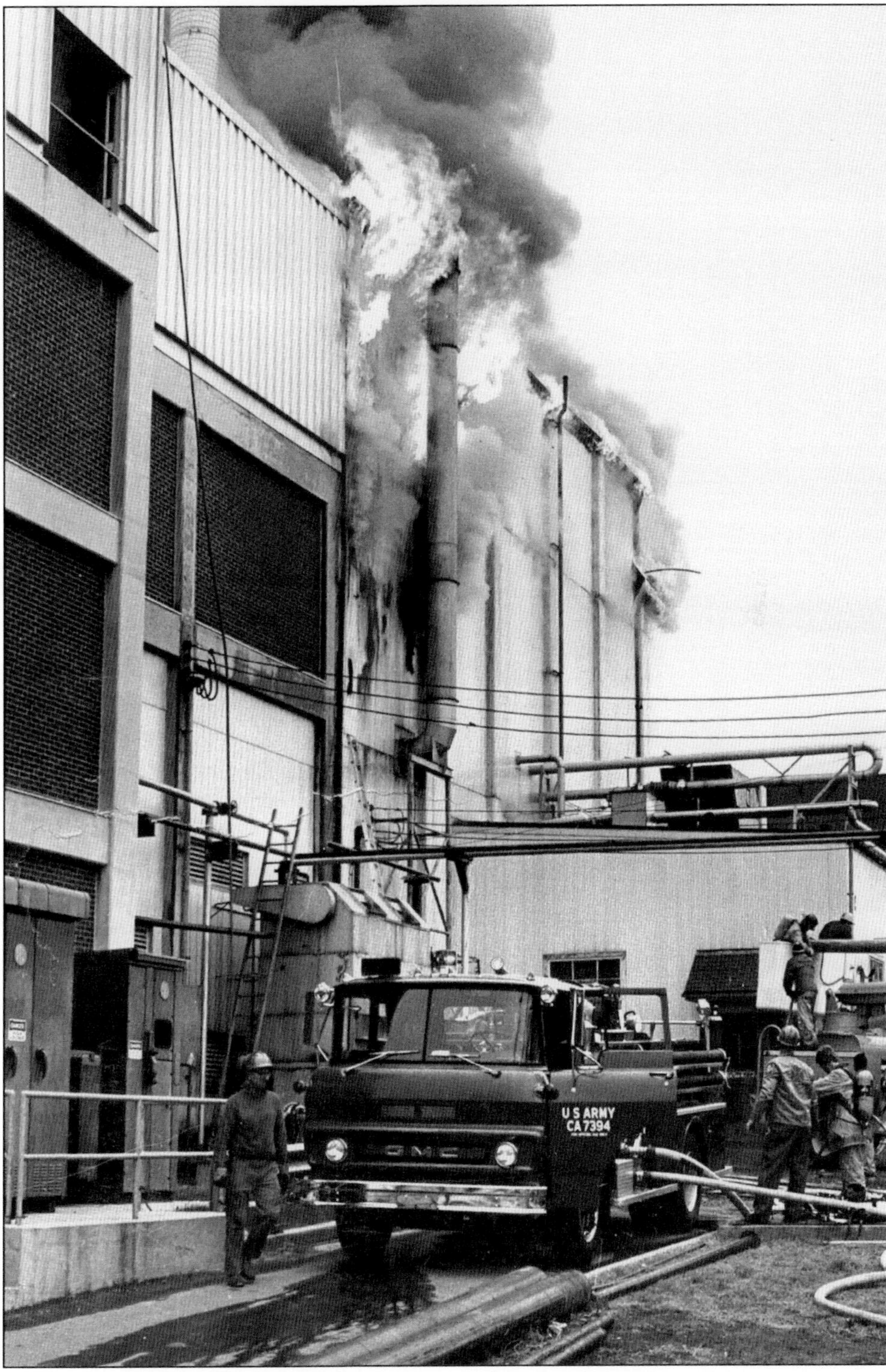

Likewise, the fire companies in Frederick have assisted Fort Detrick inside the post on major alarms. A working fire at the "8 Ball" building in the mid-1970s required a full second-alarm assignment from the Frederick companies. (Courtesy Fort Detrick Fire Department.)

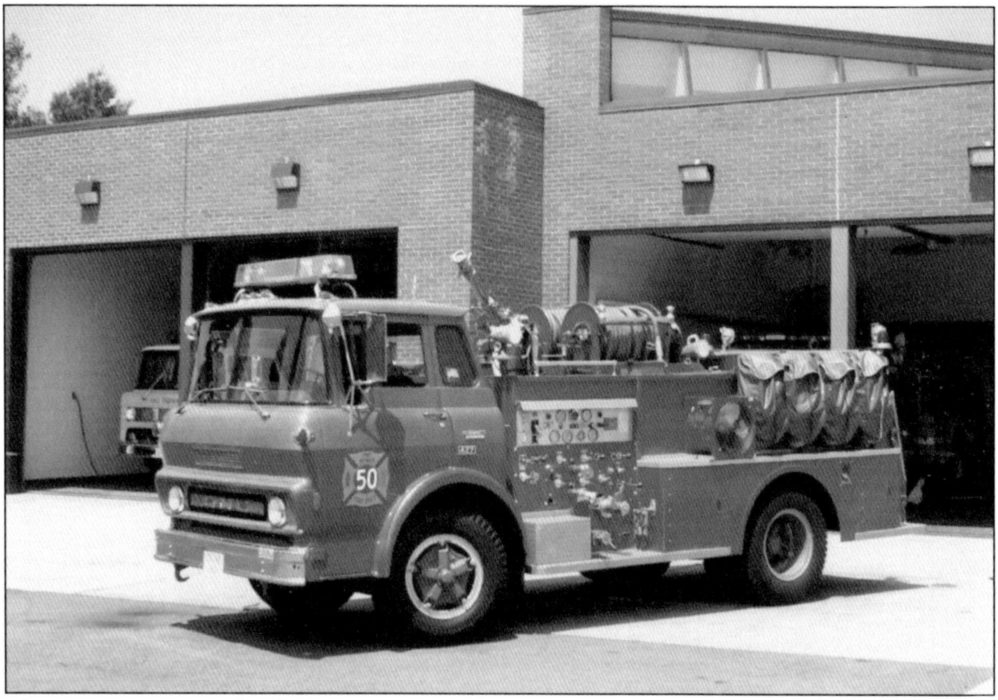

This GMC Oren was the first commercial pumper in service at Fort Detrick. Prior to this unit, military surplus engines were used by the army base. (Courtesy Warren Jenkins.)

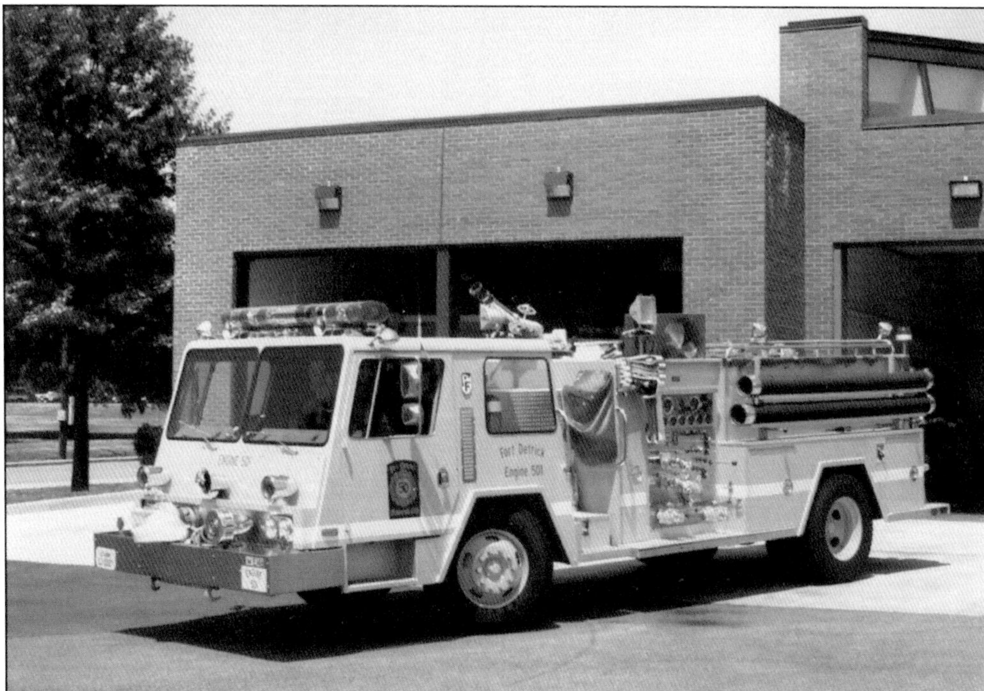

This unusual looking PenFab engine responded as Engine 501 at the Fort Detrick Fire Department. (Courtesy Warren Jenkins.)

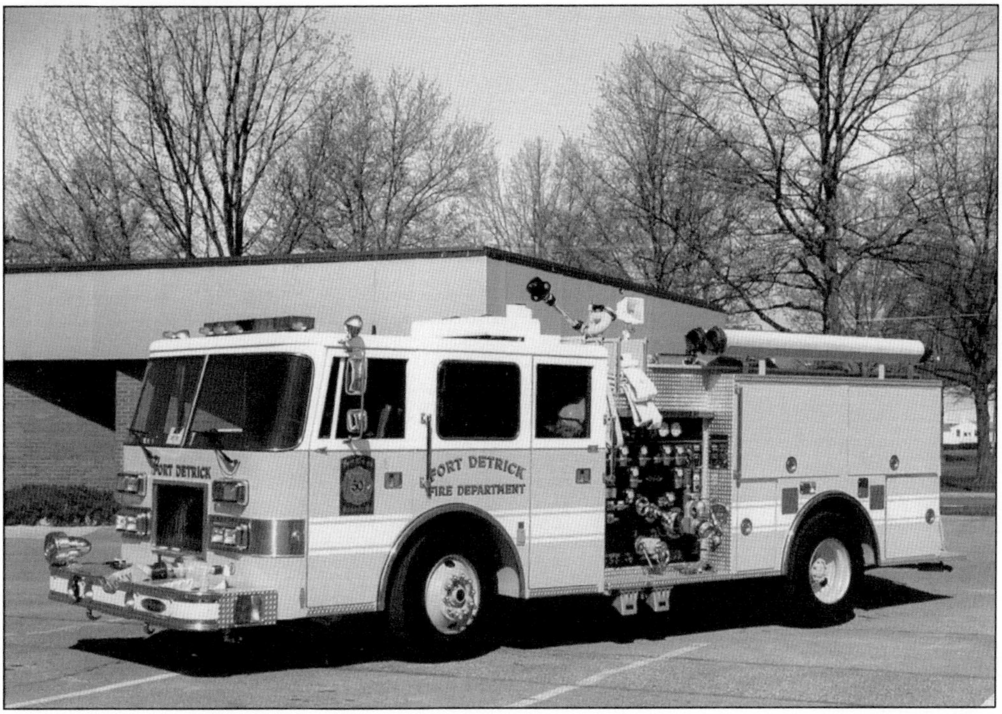

Engine 502 is a 1993 Pierce Arrow 1,250-gpm pumper. This was the last unit to be painted white over yellow. (Courtesy Warren Jenkins.)

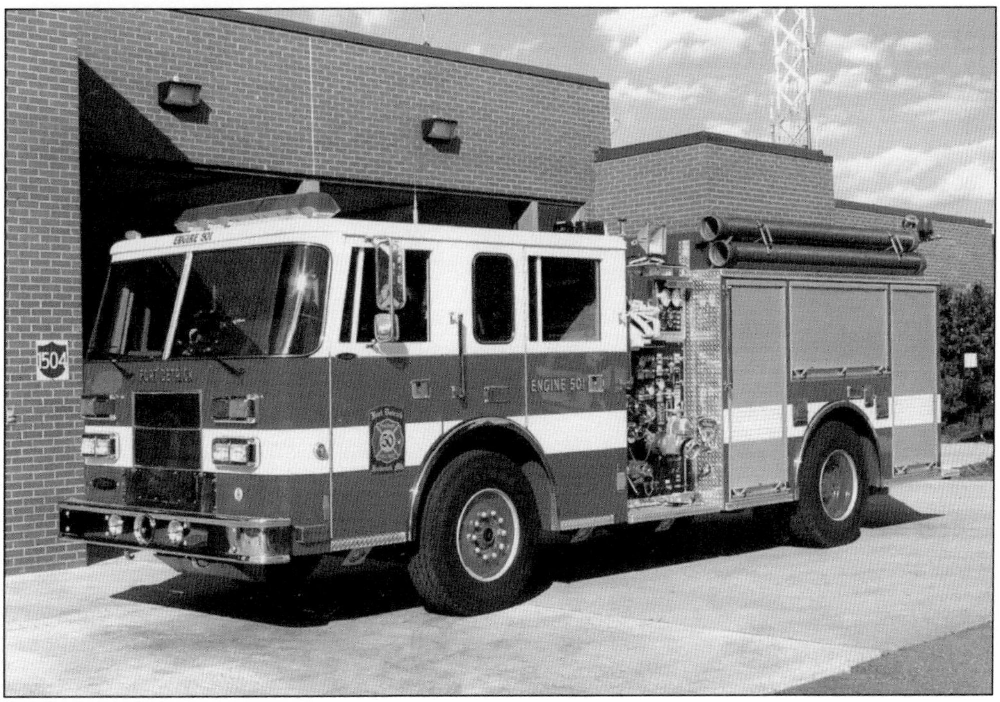

Engine 501, a 1997 Pierce Saber, is the first due engine for Fort Detrick and responds routinely off post to assist the Frederick companies.

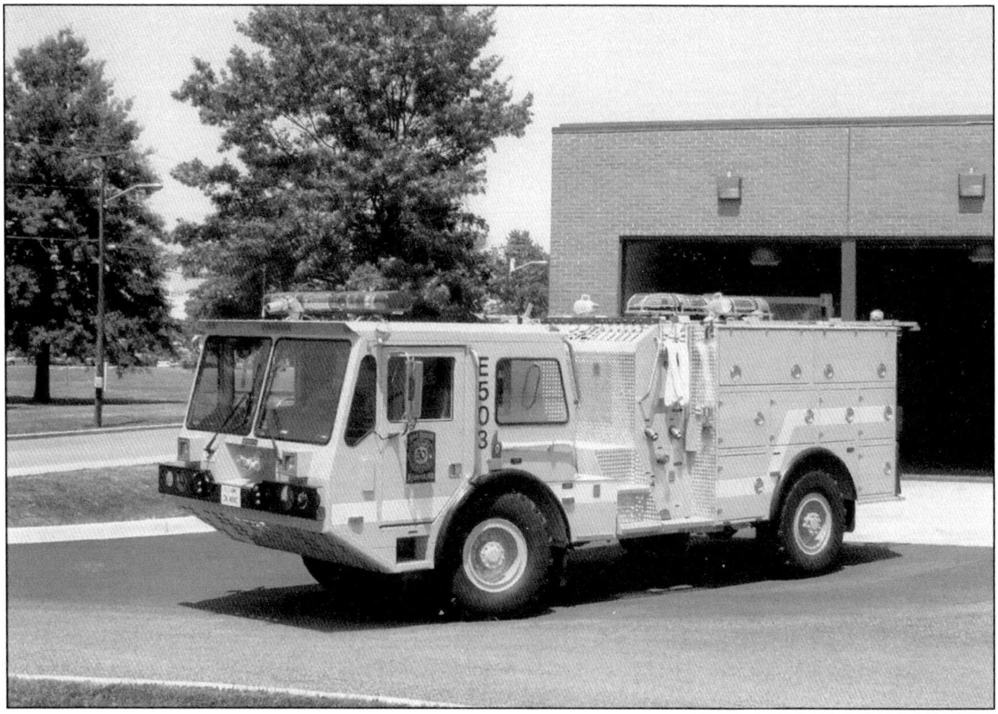

Engine 503 is an Ameritek Crash/Foam unit used for incidents requiring foam capabilities. (Courtesy Warren Jenkins.)

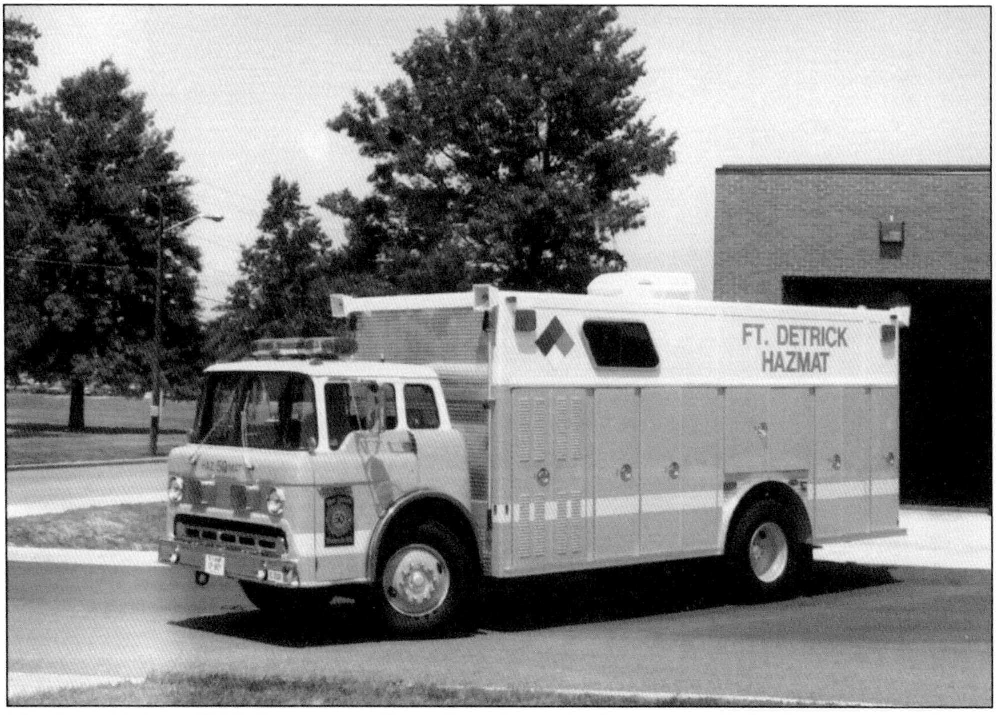

Haz-Mat 50 is a 1990 Emergency One Haz-Mat unit that responds both on post and into Frederick County. (Courtesy Warren Jenkins.)

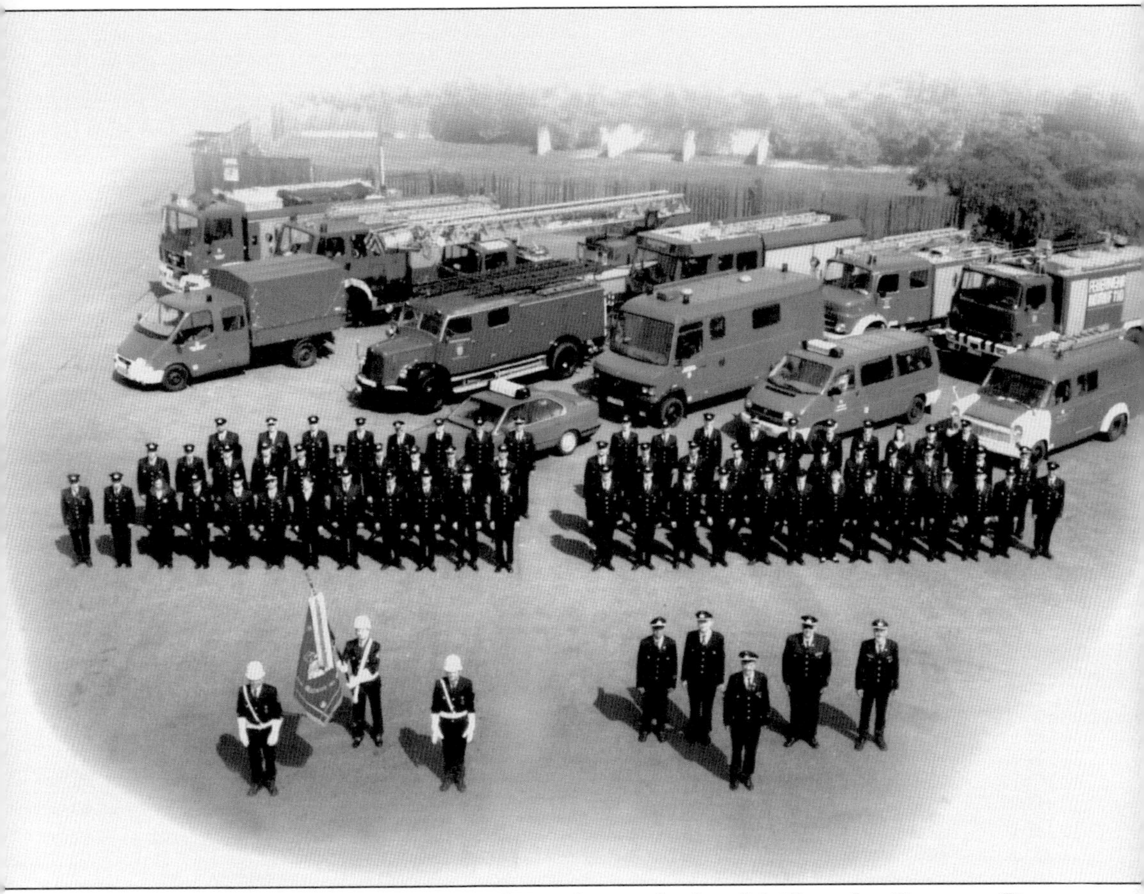

Frederick has developed a sister city relationship with the Schifferstadt, Germany Fire Department. Visitations have been conducted by both cities with firefighters visiting each country in an exchange program. (Courtesy Chip Jewell.)

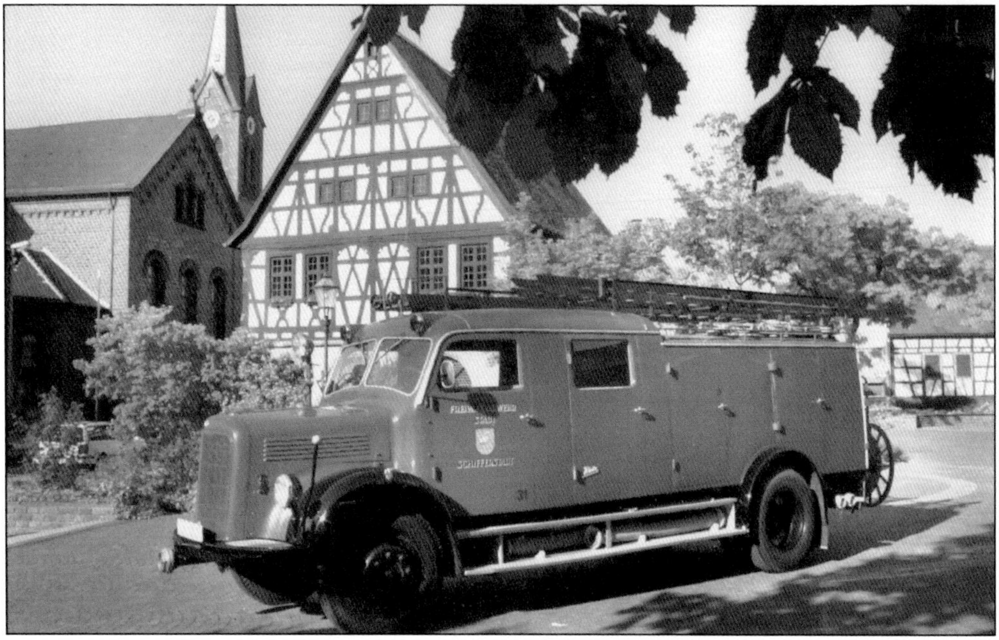

The Fuerwehr Schifferstadt still maintains this 1954 Daimler Benz engine pictured in front of the Rauthaus (city hall) in Schifferstadt, Germany. (Courtesy Chip Jewell.)

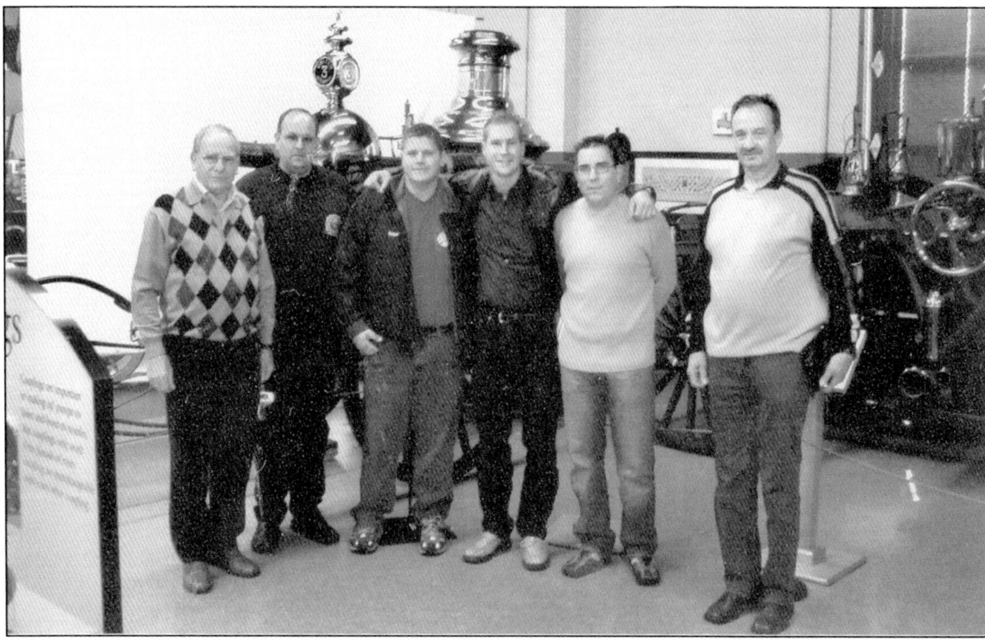

Several volunteer members of the Schifferstadt Fire Department visited Frederick in October 2003. The trip included a visit to the Fire Museum of Maryland. The honored visitors pose in front of the "Lily of the Swamp," the famed United Fire Company No. 3 steamer. Shown from left to right are Schifferstadt buergermeister (mayor) Edwin Mayer, Schifferstadt assistant fire chief Lothar Eckrich, Roger Testerman of the Walkersville Vol. Fire Company (a host family), Schifferstadt firefighters Markus Kessler and Georg Mattern, and Schifferstadt fire chief Hans Mattern. (Courtesy Chip Jewell.)

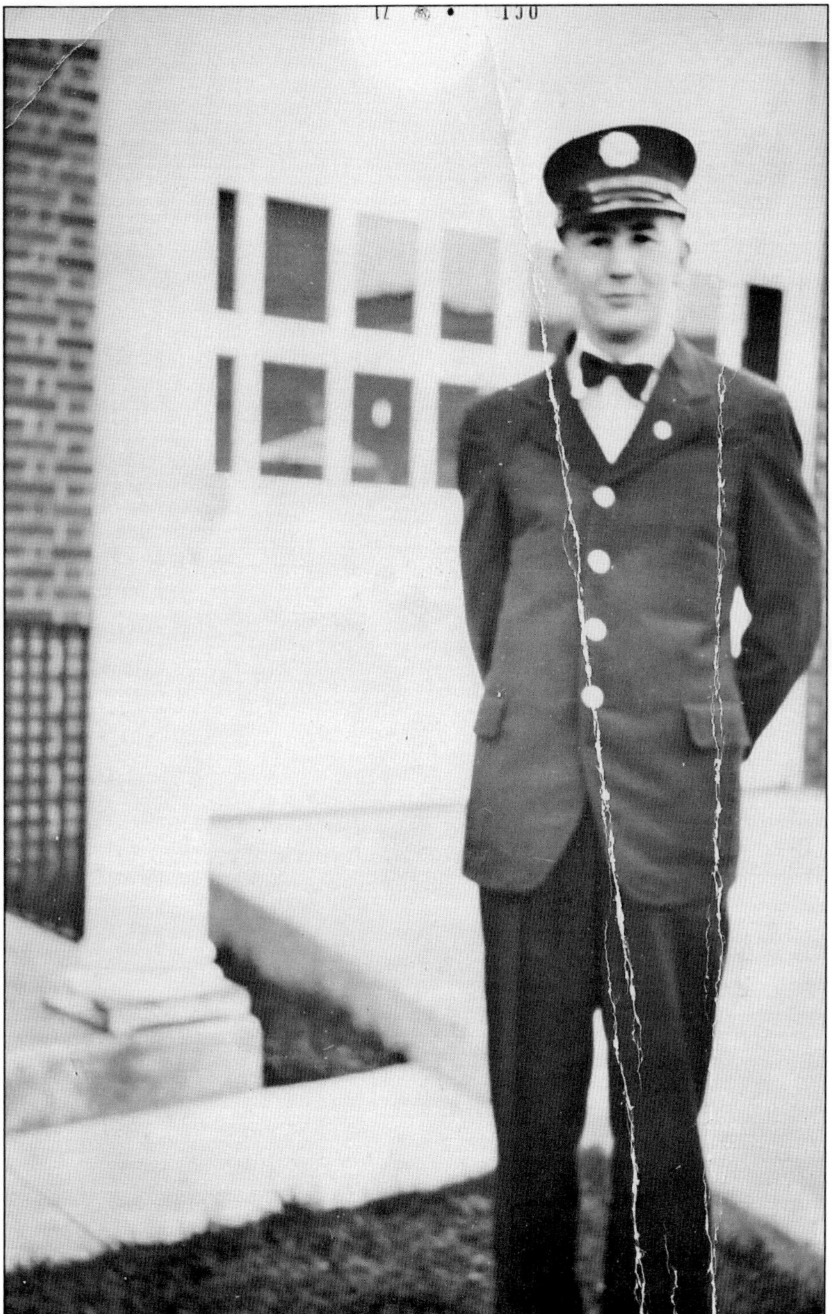

Many photos in this book were from the collection of either Charles M. Hahn or his son H. Richard Hahn. This final section is dedicated to Sgt. Charles "Mike" Hahn, a driver for the Citizens Truck Company who drove for over 30 years and an avid photographer who took many photographs used in local papers. Two sons, H. Richard "Rick" Hahn and John Hahn, and a grandson, John "Rusty" Hahn, have all followed in his footsteps and have served as career firefighters in Frederick. Their family demonstrates the tradition and commitment of many fire service families. (Courtesy JFC.)

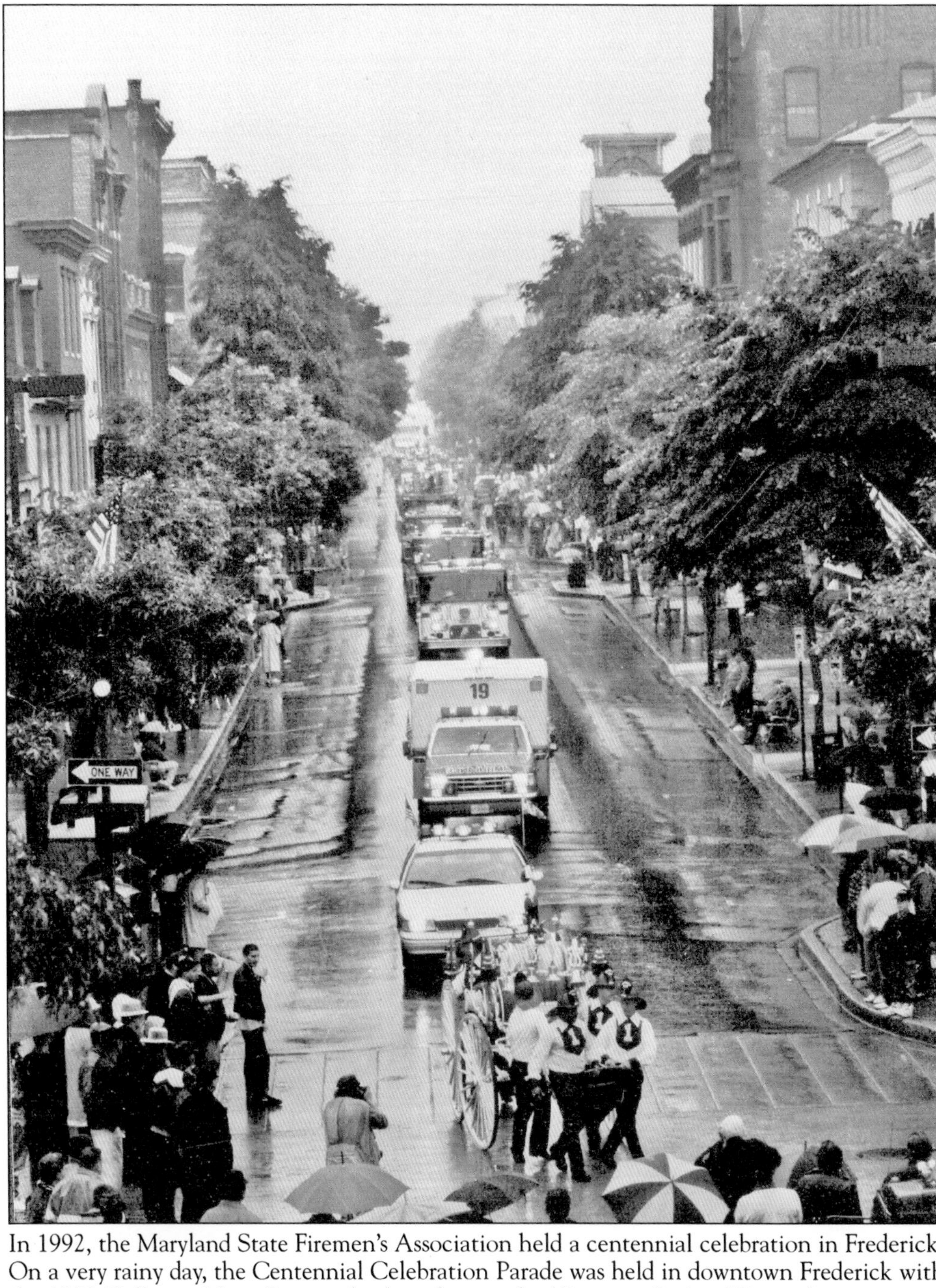

In 1992, the Maryland State Firemen's Association held a centennial celebration in Frederick. On a very rainy day, the Centennial Celebration Parade was held in downtown Frederick with the 1893 Holloway Hose Carriage leading the parade. (Courtesy Richard Yinger.)